Distributed Generation - Pioneering the Future of Decentralized Energy

Edited by Ahmed F. Zobaa and Ahmed M. Zobaa

Published in London, United Kingdom

Distributed Generation - Pioneering the Future of Decentralized Energy
http://dx.doi.org/10.5772/intechopen.1003406
Edited by Ahmed F. Zobaa and Ahmed M. Zobaa

Contributors
Faheem Mushtaq, Farshid Keynia, Francois Sekyere, George Alexandru Florea, Golamreza Memarzadeh, Hong Zhao, Isaac Prempeh, Laurentiu Constantin Lipan, Matthew E. Oliver, Patrick N. Ayambire, Prince Asabere, Tiande Mo, Willie K. Ofosu, Yong Lyu

First published in London, United Kingdom, 2025 by IntechOpen
IntechOpen is the global imprint of INTECHOPEN LIMITED, registered in England and Wales, registration number: 11086078, 167-169 Great Portland Street, London, W1W 5PF, United Kingdom

For EU product safety concerns: IN TECH d.o.o., Prolaz Marije Krucifikse Kozulić 3, 51000 Rijeka, Croatia, info@intechopen.com or visit our website at intechopen.com.

British Library Cataloguing-in-Publication Data
A catalogue record for this book is available from the British Library

Distributed Generation - Pioneering the Future of Decentralized Energy
Edited by Ahmed F. Zobaa and Ahmed M. Zobaa
p. cm.
Print ISBN 978-1-83769-742-7
Online ISBN 978-1-83769-741-0
eBook (PDF) ISBN 978-1-83769-743-4

If disposing of this product, please recycle the paper responsibly.

Meet the editors

Dr. Ahmed F. Zobaa is a distinguished scholar in electrical and power engineering with a robust academic background and extensive professional achievements. He earned his B. Sc. (Hons), M. Sc., and Ph.D. in electrical power and machines from Cairo University, Egypt, and holds a Doctoral of Science degree from Brunel University of London, U. K. Dr. Zobaa's career has spanned various roles from Instructor to Reader at notable institutions, including Cairo University and Brunel University of London. His expertise encompasses power quality, renewable energy, smart grids, energy efficiency, and lighting applications. He serves as the Executive Editor-in-Chief of the Smart Grids and Sustainable Energy Journal and holds prestigious memberships and fellowships in numerous engineering and scientific societies, underscoring his contributions to energy engineering and higher education.

Dr. Ahmed M. Zobaa is an Assistant Professor of Electrical Power and Machines at Cairo University, Egypt. He received his B. Sc. (Hons), M. Sc., and Ph.D. degrees from Cairo University in 2013, 2018, and 2023, respectively. He served as a Teaching Assistant (2013–2018), Assistant Lecturer (2018–2023), and was appointed Assistant Professor in January 2024. From 2014 to 2023, he also held a part-time Teaching Assistant position at the American University in Cairo (AUC). His research interests focus on power quality, photovoltaics, and renewable energy. Dr. Zobaa actively contributes to advancing sustainable energy technologies and engineering education in Egypt.

Contents

Preface

The evolution of power systems is increasingly marked by a shift away from centralized architectures toward more decentralized, intelligent, and environmentally sustainable energy networks. This transition is not only a response to the global imperative of carbon reduction but also a technological and economic opportunity to build resilient and efficient energy systems for the future. This edited volume, *Distributed Generation – Pioneering the Future of Decentralized Energy*, explores this multifaceted transformation by bringing together research that spans optimization, uncertainty management, hydrogen and biomass-based renewables, economic considerations, and power quality in modern distributed generation environments.

The first section, *Distributed Generation Technologies*, opens with Chapter 1, *A Comparative Analysis of Optimization Techniques for Distributed Generation (DG) in a 33 kV Radial Distribution System*. This chapter, by Prince Asabere, Isaac Prempeh, Francois Sekyere, Willie K. Ofosu, and Patrick N. Ayambire, presents a comparative evaluation of three widely used optimization algorithms—Artificial Bee Colony, Differential Evolution, and Particle Swarm Optimization—for improving voltage profiles, reducing power losses, and minimizing system costs in a 33 kV radial distribution network. The results of the study provide valuable insights for planners and engineers working on the integration of renewable distributed energy sources, particularly in developing regions.

Chapter 2, *Decreasing DG Generation Uncertainty Prediction in Energy System Management*, authored by Farshid Keynia and Golamreza Memarzadeh, delves into the challenges associated with forecasting the output of intermittent renewable resources, especially wind and solar power. The chapter highlights the role of advanced prediction techniques, including deep learning models and hybrid approaches, to improve the reliability of forecasts and enhance the operational and economic performance of power systems influenced by uncertain renewable generation.

In the second section, *Renewable Energy Sources and Applications*, Chapter 3 is titled *Insight into Hydrogen: The Future's Most Promising Energy Carrier in Preparation, Storage, Transportation, and Applications*, contributed by Tiande Mo, Yong Lyu, Faheem Mushtaq and Hong Zhao. This comprehensive chapter investigates the role of hydrogen as a clean energy vector, examining methods of production—including green hydrogen via electrolysis—alongside transportation strategies and applications in fuel cells, synthetic fuels, and other emerging sectors. The discussion emphasizes hydrogen's potential in supporting decarbonization across multiple industries.

The final section, *System Integration and Economic Analysis*, begins with Chapter 4, *The Economics of Integrating Distributed Energy Resources into the Electricity System*, by Matthew E. Oliver. This chapter offers a comprehensive overview of the economic dynamics associated with distributed energy resources, focusing on intermittency, merit-order effects, price volatility, and system reliability. It also considers market mechanisms and policy frameworks that influence the long-term adoption and integration of DERs within restructured electricity markets.

Chapter 5, *Power Quality Monitoring and Equipment in Distribution Microgrids: Insights from a Case Study*, by Laurentiu Constantin Lipan and George Alexandru Florea, closes the volume with a case-based analysis of microgrid operation. The chapter evaluates both technical and policy considerations in the integration of diesel and photovoltaic systems, emphasizing the importance of power quality monitoring, automatic switching, and energy management in remote and industrial settings. It further highlights the relevance of tools such as Reclosers, SCADA systems, and Master-Slave configurations in achieving a stable and sustainable energy supply.

This volume is designed to serve researchers, engineers, policymakers, and utility stakeholders seeking a deeper understanding of distributed energy systems. Each chapter presents practical case studies, methodological rigor, or forward-looking perspectives that collectively contribute to the global dialogue on sustainable energy innovation. As distributed generation becomes a central pillar of energy transition strategies, the research gathered in this volume will continue to inform technical development, economic planning, and policy design.

The editors wish to express their sincere appreciation to all contributing authors for their thoughtful and rigorous work. We also thank the reviewers for their insightful comments and the editorial team for their tireless support throughout the publication process. We hope that this book will serve as both a resource and a catalyst for advancing decentralized energy systems in the years to come.

Ahmed F. Zobaa
Electronic and Electrical Engineering Department,
Brunel University of London,
Uxbridge, United Kingdom

Ahmed M. Zobaa
Electrical Power Engineering Department,
Faculty of Engineering,
Cairo University,
Giza, Egypt

Section 1

Distributed Generation Technologies

Chapter 1

A Comparative Analysis of Optimization Techniques for Distributed Generation (DG) in a 33 kV Radial Distribution System

Prince Asabere, Isaac Prempeh, Francois Sekyere, Willie K. Ofosu and Patrick N. Ayambire

Abstract

The incorporation of renewable distributed generation (DG) stands as a pivotal component in implementing strategies directed toward mitigating power losses within electrical networks. With the continual rise in electricity demand and the accompanying maintenance expenditures, the importance of DG has become increasingly apparent. The effectiveness of reducing power losses, improving voltage profiles, and lowering system expenses was evaluated in this study by comparing the techniques of Artificial Bee Colony (ABC), differential evolution (DE), and particle swarm optimization (PSO). Among these methods, ABC emerged as the most effective, showing substantial potential in enhancing DG performance within a 33 kV distribution system, resulting in a 98.97% reduction in system costs. The findings of this analysis provide appreciated insights for guiding future research efforts in optimizing electrical systems.

Keywords: renewable energy technologies, power loss, electrical network, PSO, ABC, DE, system cost

1. Introduction

Electricity's pivotal role in modern activities has fueled a growing interest in alternative energy sources for power generation. To address emissions and environmental issues, the energy sector has turned to sustainable generation units like distributed generation (DG) or renewable energy technologies (RET). These innovative approaches enhance grid reliability and reduce losses, aligning with global emissions concerns and prompting a shift to sustainable technologies among electricity providers [1–4]. General Electric [2] has reported a global shift away from traditional electricity systems based mainly on large, centralized generation, transmission, and distribution (T&D) technologies toward a more distributed, digitally enhanced, and low-carbon technology-driven system. This transformation combines traditional and emerging technologies, both physical and digital, large, and small, to establish a new power network for the twenty-first century [5].

There is a global trend toward utilizing electricity generated from renewable sources [6, 7]. This shift toward renewable energy technology is fueled by its sustainable, abundant, and environmentally friendly characteristics. According to the Global Status Report published in 2021 [8], the global total renewable power capacity reached 1246 GW in 2018, with China leading at 404 GW, followed by the United States of America at 180 GW, and Germany at 113 GW. These countries are pioneering renewable energy adoption compared to the rest of the world combined.

In electricity distribution networks, ensuring power quality and stability is vital to meet increasing demand. Renewable energy technologies (RETs) are integral for addressing power quality concerns like voltage fluctuations and harmonics. Efficient RET management enhances voltage regulation, system stability, and reduces environmental impact, emphasizing the need to optimize RET operations for improved efficiency and reduced energy losses [1, 6, 9, 10].

To achieve optimal control strategies for RETs, various optimization techniques are employed. These include bio-inspired algorithms such as the Cuckoo Search Algorithm (CSA) [11, 12], Ant Colony Algorithm (ACA) [13], Gray Wolf Optimizer (GWO) [14], Analytical Coyote Optimization (ACO) [15, 16], Ant Lion Optimization Algorithm (ALOA) [17], Particle Swarm Optimization (PSO) [18, 19], Differential Evolution (DE) [20, 21], Harmony Search Algorithm (HAS) [22], Artificial Bee Colony (ABC) [23], among others. While each method offers distinct advantages and procedures for identifying optimal solutions, limited research has been conducted on the comparative performance of PSO, ABC, and DE on real networks that can be applied to evaluate distribution systems in other developing countries, aiding in their proper planning and optimization. The rationale behind the selection of PSO, ABC, and DE lies in their established effectiveness and widespread adoption in power system optimization studies. These techniques have demonstrated notable success in addressing similar optimization challenges, particularly in radial distribution systems.

Therefore, this chapter aims to conduct a comparative analysis of DE, ABC, and PSO optimization techniques for RETs in a 33 kV radial distribution system in Ghana, providing valuable insights that can be directly applied to similar systems in other developing countries. The study investigates three optimization techniques to identify the most effective approach for deploying RET in distribution systems. The evaluation process considers the techniques' ability to improve voltage profile, reduce power losses, and minimize system cost. By comparing these methods against these predefined performance criteria, valuable insights are gained to guide selecting the most suitable approach for RET integration.

2. Renewable energy technologies landscape

More than one billion people worldwide lack access to electricity, and close to three billion rely on biomass, such as wood, for cooking [24, 25]. Sub-Saharan Africa (SSA) and developing Asia account for over 95% of those without electricity. In Africa, the renewable energy sector comprises solar photovoltaic, wind, hydro, geothermal, and biomass sources. The utilization of a combination of these resources depends mainly on their availability in the beneficiary country's location. In 2019, the renewable energy capacity in SSA was estimated to be around 50 GW [26]. Hydro-generation plants contributed the largest share of total power production at 35.7 GW.

Non-hydro sources such as wind, geothermal, biomass, and solar accounted for less than 13.8 GW [26, 27].

The solar potential in Africa ranks among the highest globally, yet only a small fraction of it is harnessed for electricity generation. According to the General Electric report on energy generation in SSA, solar generation in this region represents just 1% of global solar generation [26–28]. Ref. [27] noted that most areas in SSA receive over 2000 kWh of solar radiation annually. However, only 8% of these renewable sources are currently utilized for electricity generation. Recognizing this significant potential, countries in the sub-region have outlined plans to venture into the RET sector as electricity demand continues to rise annually [26].

To harness the potential of RET and meet the power demand in the sub-region, countries have established the West African Power Pool [29, 30]. These strategies have emerged due to the interdependence of regional countries and their shared goal of promoting stable, reliable power at affordable rates. Consequently, the planning and outlook for Renewable Energy have followed the formal adoption of two major regional policy developments: the 2011/12 West African Power Pool (WAPP) Master Plan and the ECOWAS Renewable Energy Policy (EREP). These policies aim to increase the region's overall share of renewable energy in the electricity generation mix to 23% by 2020 and 31% by 2030 [29, 31].

3. Distribution system configuration (33 kV)

This study optimizes renewable energy technologies (RET) deployment in a 33 kV radial distribution system in the Ashanti region, Ghana. The system has 49 branches and 50 bus bars that stretch 574.4 kilometers. For RET integration, the study compares DE, ABC, and PSO optimization methods. The ECG 33 kV network diagram is shown in **Figure 1** [32].

Each technique is evaluated based on its ability to determine the best placement and sizing of RET devices for achieving voltage profile improvement (VPI), power loss reduction (PLR), and minimized system cost (SC). The predefined performance metrics, including VPI, PLR, and SC, act as evaluation criteria for comparing the effectiveness of the optimization techniques. This comparison allows for the selection of the most effective optimization approach for RET deployment in the system.

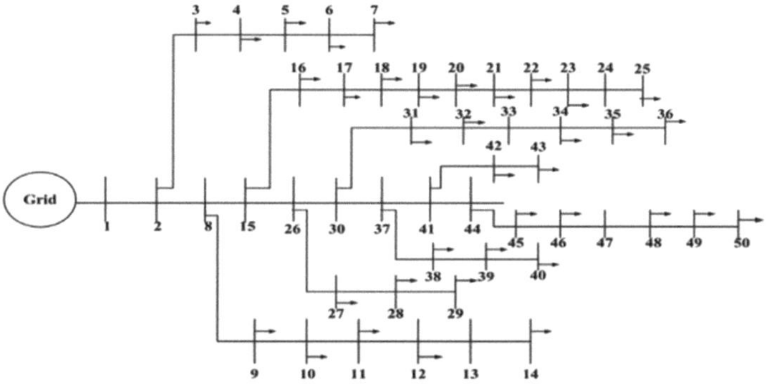

Figure 1.
A single-line diagram of 33 kV distribution network of Ashanti region.

4. Problem formulation and constraints

When Renewable Energy Technologies (RET) are linked to a bus within the grid, they provide the ability to manage voltage regulation, frequency, and reactive power. This capability has an impact on the voltage profile of the grid, as described by Ogunsina et al. [13] and Abdallah et al. [15]. The primary aim of integrating RET into the distribution system is to uphold the stability and dependability of the system, particularly during variations in demand or unforeseen circumstances. Therefore, the initial objective function (power loss reduction) can be expressed as

$$P_{LR} = f_1 \tag{1}$$

$$f_i = P_{loss} = \sum_{i=1}^{n} R_i I_i^2 \tag{2}$$

Where
f_1 is the first objective function related to power losses,
P_{loss} *is* active power loss,
I_i is line i current,
R_i is the resistance of ith line,
n is the number of system busses.
The total power loss regulation can be formulated as

$$P_{TL}\ (\%) = \frac{P_{tlb(RET)} - P_{tla(RET)}}{P_{tla(RET)}} * 100 \tag{3}$$

Where
$P_{TL}\ (\%)$ is the total power loss regulation in percentage,
$P_{tlb(RET)}$ is the total power loss before RET integration,
$P_{tla(RET)}$ is the total power loss after RET integration in the system.
The second objective function (voltage profile improvement) of the system can be expressed as,

$$f_2 = V_{PI} = \sum_{i=1}^{n} (V_n - V_i)^2 \tag{4}$$

Where
V_{PI} is the voltage profile improvement,
V_n is the system's nominal voltage, which is 1 per unit (1p.u.),
V_i is the voltage on the ith bus,
n represents the number of busses on a radial system.
The third objective function (system cost) can be written as [32]

$$f_3 = T_{sc} = kz\left(P_{tlb(RET)}\right) - kz\left(T * P_{tla(RET)}\right) - \left(kx * RET_{\frac{cost}{year}}\right) \tag{5}$$

Here, $T, kx,$ *and* kz signify the hour/year, duration proportion, and cost of energy losses, respectively.

5. Load flow analysis

A Load Flow Analysis is an initial step in assessing interconnected power systems, providing essential insights not only during system design and expansion but also during operational phases for control and economic scheduling. It aims to accurately determine steady-state voltages, voltage angles, and power flows within a network, crucial for maintaining system stability and reliability [33]. Load flow methods like Gauss-Seidel, Newton-Raphson, Fast-decoupled, and Back/Forward Sweep (BFS) approaches are commonly used for load flow analysis, each with distinct characteristics and computational requirements [34–36].

The Gauss-Seidel method, known for its sequential displacement approach, updates values iteratively and converges with high sensitivity to initial values, while Newton-Raphson approximates nonlinear equations with quadratic convergence, making it more suitable for complex problems but demanding more computational resources. However, both methods face challenges in practical networks due to high R/X ratios, making backward/forward sweep (BFS) method preferred for radial networks. BFS splits equations into smaller systems, solving them sequentially for convergence, and can simulate dispersed generation like solar panels in radial networks [34–36].

The performance and resilience of Gauss-Seidel, Fast-decoupled, and Newton-Raphson methods in distribution systems can be compromised due to potentially inaccurate simplification assumptions, despite their standard application in transmission systems. Hence, a backward/forward load flow algorithm is preferred for its suitability and accuracy in distribution system analyses. The BFS load flow formulations and constraints can be found in the works of Pesaran et al. [37], Montoya et al. [38], Heidarifar et al. [39], and Ghatak and Mukherjee [40].

6. Mathematical formulation of PSO, ABC, and DE optimization

6.1 Particle swarm optimization method

Particle swarm optimization (PSO) is a metaheuristic algorithm introduced by Kennedy and Eberhart in 1995, inspired by bird flocks or fish schools. In PSO, a population of candidate solutions, called particles, moves around in the search space to find the optimal solution. Each particle represents a potential solution to the optimization problem. The movement of particles is guided by their own best-known position and the global best-known position of the entire swarm. This information is used to update the velocity and position of each particle in each iteration. The basic steps of the PSO algorithm involve initializing a population of particles randomly within the search space, evaluating the fitness of each particle using an objective function, updating the particle's velocity and position based on its own best-known position and the global best-known position of the entire swarm until a stopping criterion is met [41].

However, PSO may struggle with local optima and may require tuning of parameters such as the number of particles, the inertia weight, and the acceleration coefficients to achieve good performance. Additionally, PSO does not guarantee global optimality but rather tends to converge to a good solution in a reasonable amount of time. PSO is a powerful and widely used optimization algorithm with applications in various domains, including engineering, economics, machine learning, and biology [42–45].

The possible solutions involve two vectors that represent the position (X) and velocity (V) of a particle. Following each iteration, the particle with the best solution shares its location coordinates (P_gbu) with the rest of the swarm. The updated velocity and position of the particle indexed as (i^{th}) at a specific time (t) are expressed by Eqs. (6 and 7) respectively.

$$V_{iu}^{t+1} = w * V_{iu}^{t} + C_1 * r_1 * (P^t + X^t) + C_2 * r_2 * \left(P_{gbu}^{t} - X_{iu}^{t}\right) \tag{6}$$

$$X_{iu}^{t+1} = X_{iu}^{t} + V_{iu}^{t+1} \tag{7}$$

Here, C_1, C_2 =acceleration coefficient, $u = 1, 2 \ldots m$ (particle-searching space members), r_1, r_2 = random numbers (to introduce stochasticity, the model involves uniformly distributed random numbers in the range $[0, 1]$), $i = 1, 2 \ldots n$ (the number of particles in the swarms), V_{iu}^{t} is present speed of the particle $('i')$, V_{iu}^{t+1} is the particle's updated velocity $('i')$, P^t represents the particle's optimal position, P_{gbu} is the best value for the group of the particle, X_{iu}^{t} is Current search point of a particle, X_{iu}^{t+1} is the updated searching point of a particle, and w is inertial weight. The inertia weight function can be expressed as,

$$w = w_{max} - \frac{w_{max} - w_{min}}{t_{max}} * t \tag{8}$$

Here, w_{min} and w_{max} are the minimum and maximum inertia weights, respectively, and $t \; and \; t_{max}$ are the present and maximum iteration.

6.2 Artificial bee colony method

The Artificial Bee Colony (ABC) algorithm is a population-based optimization technique inspired by the foraging behavior of honeybees. ABC was introduced by Karaboga in 2005. It mimics the food search process of honeybees, which involves employed bees, onlooker bees, and scout bees [23]. The employed bees explore random areas, share findings with onlooker bees who choose better areas based on the information, and scout bees venture into entirely new areas if improvement stagnates, all working together until the best solution is identified. One of the key advantages of the ABC algorithm is its simplicity and ease of implementation. It is also capable of efficiently exploring complex search spaces and has been successfully applied in various optimization problems, but it tends to be more exploratory, which can sometimes delay convergence to the global optimum. The fitness function of ABC quantifies the quality of a candidate solution [23, 46, 47], which can be expressed as:

$$f_{fit} = \frac{1}{1 + p_{losses}} \tag{9}$$

Onlooker bees consider the information shared by employed bees about food sources and update their own knowledge using Eq. (10).

$$f_{ij} = f_{minj} + rand\left(f_{maxj} - f_{minj}\right) \tag{10}$$

The minimum and maximum values for the jth variable are represented as f_{minj} and f_{maxj} correspondingly. A random number (rand) is selected from the range of −2 to 2 [48]. Eq. (11) directs the onlooker bee in choosing food sources randomly.

$$p_{prob(i)} = \frac{f_{it(i)}}{\sum_k^N f_{it(g)}} \tag{11}$$

Here, $f_{it(i)}$ represents the fitness value at the chosen solution (i). The fitness value at each iteration is denoted as $f_{it(g)}$, where the variable length is g. The scout bee utilizes Eq. (11) when the newly updated solution is inferior to the previous best solution.

Therefore,

$$f_{it(i)} = \begin{cases} \frac{1}{1+f_i} \; if \; f_i \geq 0 \\ 1+abs\left(f_i\right) \; if \; f_i < 0 \end{cases} \tag{12}$$

Here, f_i denotes the cost value of the objective function.

6.3 Differential evolution method

Differential Evolution (DE) is an optimization algorithm developed by Storn and Price in 1997, widely used for solving optimization problems across different domains due to its simplicity and effectiveness [21]. In DE, a population of candidate solutions is randomly initialized within the search space, with each individual representing a potential solution. The algorithm evolves this population iteratively to search for the optimal solution [21].

DE creates new trial solutions at each iteration through several steps: first, a mutant vector is generated by adjusting an individual's position based on differences with randomly selected individuals from the population. Next, a crossover operation combines information from the mutant vector and the target individual to form a trial vector. This trial vector undergoes evaluation against the target individual based on their fitness values, and if superior, replaces the target individual in the next generation. These steps aid in the efficient exploration of the search space while balancing exploration and exploitation to reach the optimal solution.

DE's strengths lie in its robustness, simplicity, and ability to handle optimization problems with nonlinear, non-smooth, and multimodal objective functions, but sometimes get trapped in local optima, particularly in multimodal landscapes with numerous local peaks. It finds applications in engineering design, machine learning, finance, and bioinformatics. The size of the population and the dimensions of optimization variables in each individual string, which is known as the target vector (TV), are initialized, and generated randomly according to Ramli et al. [21], Mohamed [49], and Liao et al. [50].

$$TV_{iG}^{j} = round\left[TV_{mn}^{j} + rand() * \left(TV_{mx}^{j} - TV_{mn}^{j}\right)\right] \tag{13}$$

Here, j = 1, 2, …, D, i = 1, 2, …, Np, G is the generation number, and mn and mx are the minimum and maximum vectors, respectively. Eq. (14) generates a mutation vector (MV_i) for each target vector. It can be expressed as

$$MV_{iG} = TV_{bG} + F * \left(TV_{r1G} - TV_{r2G}\right) \tag{14}$$

The mutation operation involves a constant scaling factor (F) ranging from $0 \; to \; 2$, along with two random integers $(r1 \; and \; r2)$ and the best target vector (TV_b) in the population for that generation (G) based on fitness function values. After mutation, a crossover operation is performed on each pair of the target vector (TV_{iG}) and its corresponding mutant vector (MV_{iG}) to create a crossover vector $\left(CV_{jiG}\right)$ using the (Eqs. (15) and (16)).

$$CV_{jiG} = \begin{cases} MV_{jiG} \text{ if } rand\ () \leq CF \\ TV_{jiG} \text{ otherwise} \end{cases} \tag{15}$$

Here, CF is the crossover factor, and it is a user-defined parameter between the range of [0, 1]. The selection operation performed after the crossover can be expressed as:

$$TV_{iG+1} = \begin{cases} CV_{iG} \text{ if } f\ (\ CV_{iG}) \leq f\ (\ TV_{iG}) \\ TV_{iG} \text{ otherwise} \end{cases} \tag{16}$$

Here, f denotes the objectives function.

7. Results and discussion

The power system evaluation utilized DE, PSO, and ABC methods, incorporating RET integration as the primary control factor. Optimal system performance hinged on RET placement and size. This study investigated seven cases: the Base case without RET, Case 1 with one RET, Case 2 with two RETs, Case 3 with three RETs, Case 4 with four RETs, Case 5 with five RETs, and Case 6 with six RETs. Each case utilized DE, PSO, and ABC methods except the Base case. The impact of RET integration was assessed using Backward/Forward Sweep power flow analysis by examining active power losses, voltage profile, and system cost. Effectiveness was determined by comparing integrated and non-integrated RET results. MATLAB software facilitated the optimization process for RET integration on the 33 kV distribution system.

7.1 PSO method results

7.1.1 Power loss reduction (PLR)

Active power loss results were presented for Case 1 through Case 6, highlighting total active losses and the percentage reduction in active losses. These parameters were visualized in **Figure 2**, excluding the total loss of the Base case (14.53 MW), which exceeded that of the optimal cases.

When comparing power loss reductions among the six cases excluding the original system (Base case), **Figure 2** exhibits notable differences. Cases 6 and 5 demonstrate total active power losses of 0.17 MW each, with reductions of 14.36 MW and 14.36 MW, resulting in percentage reductions of 98.83% and 98.83%. In contrast, Cases 4 and 3 show total active losses of 0.23 MW and 0.2 MW, achieving reductions of 14.30 MW and 14.33 MW, leading to percentage reductions of 98.42% and 98.62%, respectively. For Case 2 and Case 1, total active losses are 0.21 MW and 0.25 MW, with power loss reductions reaching 14.32 MW and 14.28 MW and percentage reductions of 98.56% and 98.28%, respectively. The results indicate that Case 5 and Case 6 outperform other cases in the PSO method. **Figure 2** visually depicts the total active and percentage reduction in losses (PLR). Optimal active loss values are attained through strategic placement and sizing of RET, as detailed in **Table 1**.

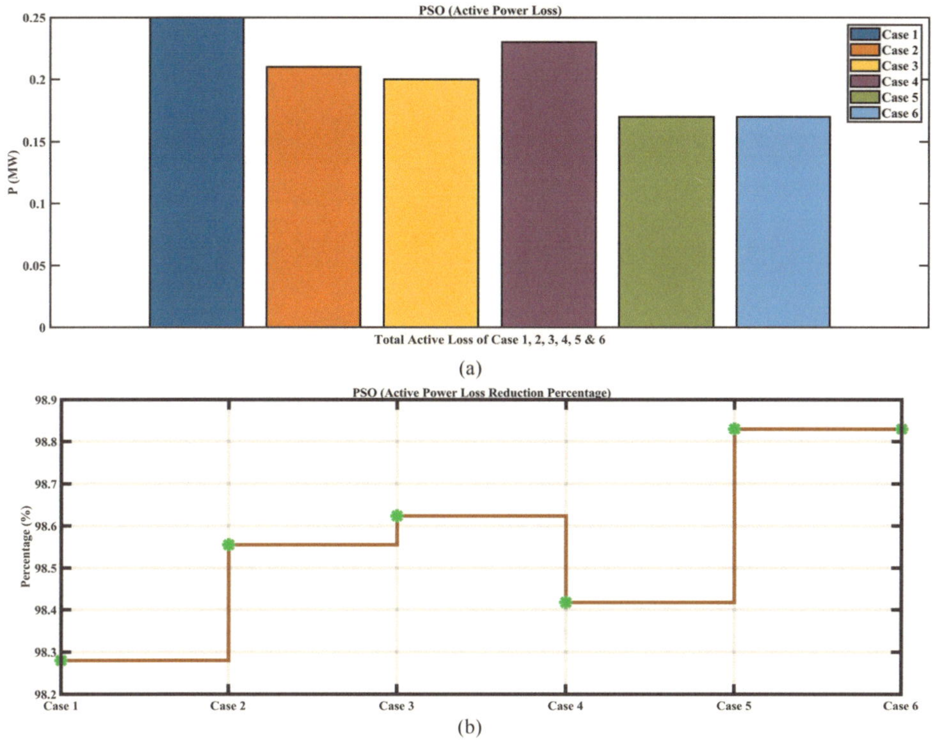

Figure 2.
Total active loss (a) and percentage reduction of active loss (b) of the PSO cases.

7.1.2 Voltage profile improvement (VPI)

Figure 3 presents the results of Voltage Profile Improvement (VPI) from the Base case to Case 6, assessing the rise in voltage profile values per unit (p.u).

It indicates in **Figure 3** that increasing the number of renewable distributed generators (RETs) leads to an enhancement in the voltage profile of the distribution system. Case 6, with six RETs, shows the most substantial improvement, especially with busses 44 to 50. This suggests that strategic placement and sizing of RETs can effectively improve the voltage profile in a distribution system.

7.1.3 System cost (SC)

The purpose of this study segment was to assess the cost-effectiveness of Cases 1, 2, 3, 4, 5, and 6. This evaluation involved analyzing both the system cost and the percentage reduction in cost for each case. **Figure 4** illustrates the system cost and the corresponding percentage reduction.

The data provided in **Figure 4** offers insights into the system cost and the corresponding percentage of cost reduction for various cases within a power system. The Base case, with a system cost of $2,441,040.00/year, serves as the reference point for evaluating cost reduction across the subsequent cases. Cases 5 and 6 display a cost reduction percentage of 98.83% for each case, marking a higher reduction than all the cases. This remarkable reduction underscores the significant cost-saving potential of integrating RET into the power distribution system, emphasizing the importance of considering such solutions in modern energy infrastructure planning and management.

Parameters	Base case	Case 1	Case 2	Case 3	Case 4	Case 5	Case 6
PSO							
Minimum bus voltage (p.u)	0.817	0.993	0.993	0.994	0.994	0.994	0.996
Maximum bus voltage (p.u)	1	1	1	1	1	1	1
DG power injection (kW) and location	—	53 at bus 19	58 at bus 32 53 at bus 19	58 at bus 32 52 at bus 19 26 at bus 14	29 at bus 39 58 at bus 32 13 at bus 43 38 at bus 46	19 at bus 47 142 at bus 44 44 at bus 9 47 at bus 19 51 at bus 32	12 at bus 47 28 at bus 39 59 at bus 32 45 at bus 19 144 at bus 15 26 at bus 14
Total active loss (MW)	14.53	0.25	0.21	0.2	0.23	0.17	0.17
Active reduction percentage	—	98.28%	98.56%	98.62%	98.42%	98.83%	98.83%
Total system cost ($/year)	2,441,040.00	42,000.00	35,280.00	33,600.00	38,640.00	28,560.00	28,560.00
Net saving ($/year)	—	2,399,040.00	2,405,760.00	2,407,440.00	2,402,400.00	2,412,480.00	2,412,480.00
System cost reduction percentage	—	98.28%	98.55%	98.62%	98.42%	98.83%	98.83%
ABC							
Minimum bus voltage (p.u)	0.817	0.994	0.996	0.995	0.996	0.996	0.996
Maximum bus voltage (p.u)	1	1	1	1	1	1	1
DG power injection (kW) and location	—	53 at bus 19	26 at bus 14 52 at bus 19	24 at bus 14 67 at bus 18 67 at bus 32	41 at bus 13 45 at bus 19 16 at bus 40 20 at bus 35	32 at bus 14 2 at bus 38 24 at bus 5 42 at bus 20 60 at bus 3	0 at bus 37 58 at bus 19 123 at 3 23 at bus 33 0 at bus 27 26 at bus 14
Total active loss (MW)	14.53	0.2	0.19	0.16	0.16	0.17	0.15
Active reduction percentage	—	98.62%	98.69%	98.9%	98.9%	98.83%	98.97%
Total system cost ($/year)	2,441,040.00	33,600.00	31,920.00	26,880.00	26,880.00	28,560.00	25,200.00
Net saving ($/year)	—	2,407,440.00	2,409,120.00	2,414,160.00	2,414,160.00	2,412,480.00	2,415,840.00

Parameters	Base case	Case 1	Case 2	Case 3	Case 4	Case 5	Case 6
System cost reduction percentage	—	98.62%	98.69%	98.9%	98.9%	98.83%	98.97%
DE							
Minimum bus voltage (p.u)	0.817	0.994	0.994	0.996	0.994	0.994	0.994
Maximum bus voltage (p.u)	1	1	1	1	1	1	1
DG power injection (kW) and location	—	52 at bus 19	73 at bus 3 47 at bus 20	55 at bus 3 85 at bus 19 31 at bus 14	0 at bus 2 57 at bus 19 46 at bus 32 0 at bus 50	44 at bus 11 31 at bus 3 64 at bus 20 0 at bus 2 0 at bus 4	0 at bus 50 0 at bus 2 52 at bus 19 0 at bus 49 0 at bus 47 67 at bus 3
Total active loss (MW)	14.53	0.2	0.17	0.19	0.17	0.18	0.17
Active reduction percentage	—	98.62%	98.83%	98.69%	98.83%	98.76%	98.83%
Total system cost ($/year)	2,441,040.00	33,600.00	28,560.00	31,920.00	28,560.00	30,240.00	28,560.00
Net saving ($/year)	—	2,407,440.00	2,412,480.00	2,409,120.00	2,412,480.00	2,410,800.00	2,412,480.00
System cost reduction percentage	—	98.62%	98.83%	98.69%	98.83%	98.76%	98.83%

Table 1.
The summary results of all cases.

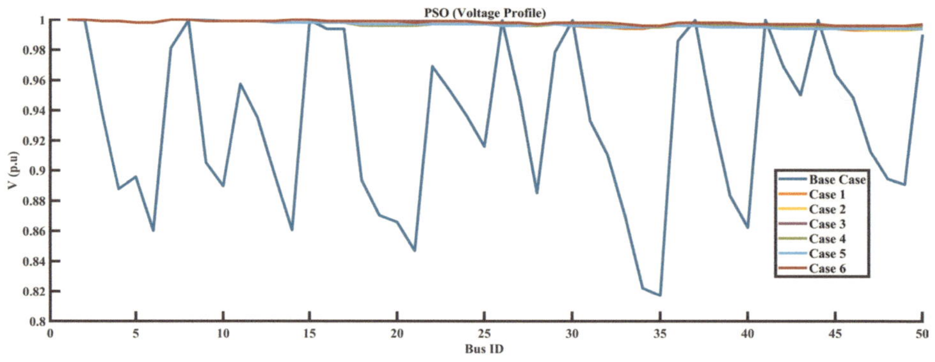

Figure 3.
Voltage profile improvement of all PSO cases.

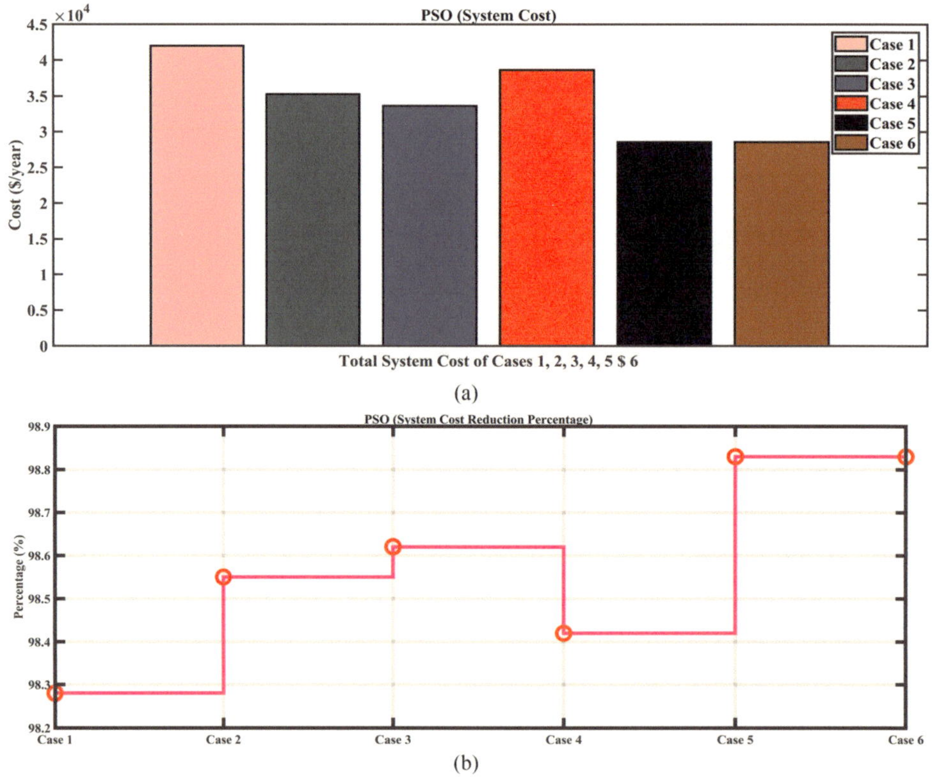

Figure 4.
Total system cost (a) and percentage reduction of system cost (b) of the PSO cases.

7.2 ABC method results

7.2.1 Power loss reduction (PLR)

Figure 5 illustrates the variations in active power losses and the related percentage reduction among different ABC cases within a power system, respectively. The Base Case, with an initial loss of 14.53 MW, serves as the reference case without any

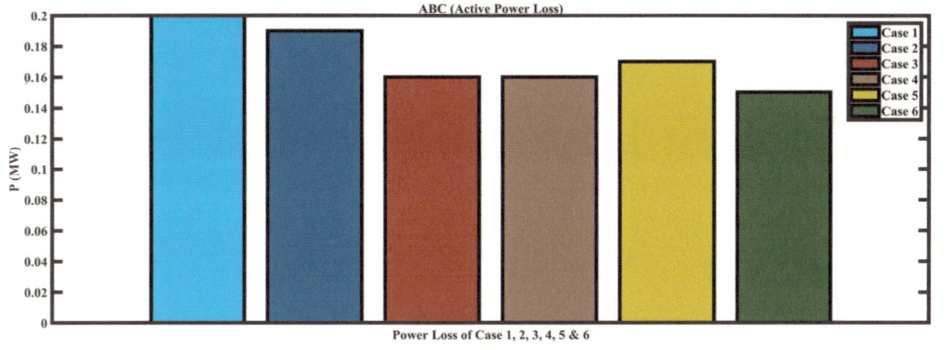

(a)

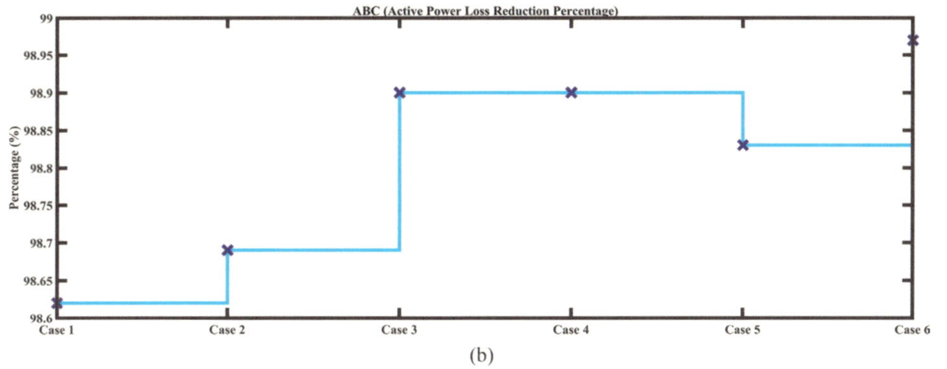

(b)

Figure 5.
Total active loss (a) and percentage reduction of active loss (b) of the ABC cases.

optimization and is excluded from **Figure 5** as it exhibits the highest active power loss among all cases. Case 1 (0.2 MW) exhibits a notable decrease in active power losses compared to the base case, showcasing an implemented improvement that results in lower losses with a percentage of 98.69%. Similarly, Case 2 (0.19 MW) also displays reduced active power losses, albeit slightly less than Case 1. Case 3 (0.16 MW) demonstrates further enhancement with even lower losses compared to Cases 1 and 2, indicating a more effective optimization approach. Likewise, Case 4 (0.16 MW) maintains consistent performance in reducing losses.

Although Case 5 (0.17 MW) shows a slight increase in losses compared to Cases 3 and 4, it still maintains significantly lower losses than the base case. Case 6 (0.15 MW) illustrates the lowest active power losses among all cases, reflecting a highly optimized or improved system configuration. According to **Figure 5**, the optimization measures implemented in Cases 1 to 6 have led to substantial reductions in active power losses compared to the base case. Case 6, with a percentage of 98.97%, shows the most notable improvement. This reduction in losses can result in improved system efficiency, cost savings, and overall enhanced performance.

7.2.2 Voltage profile improvement (VPI)

Figure 6 summarizes the voltage profiles (per unit) of various busses across different cases in an ABC method. There is a noticeable improvement in voltage profiles in Cases 1 to 6 compared to the Base case.

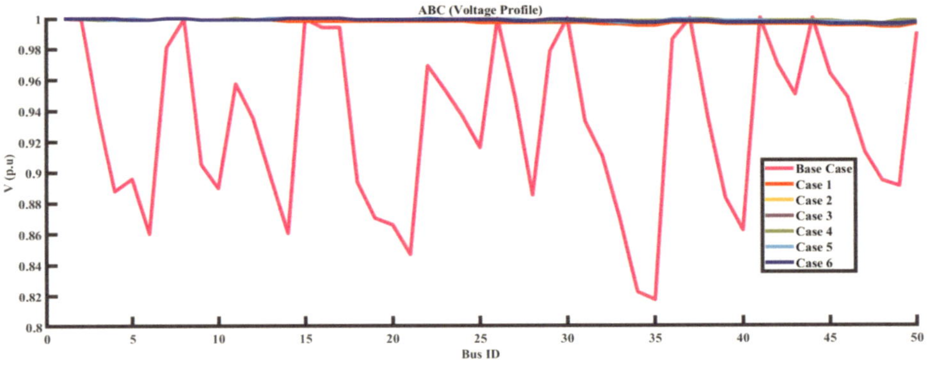

Figure 6.
Voltage profile improvement of all ABC cases.

Case 4 exhibits a good voltage profile, especially starting from bus 43 to 50, compared to other cases. This is evident from the fact that the voltage profile values at these busses in Case 4 are consistently close to 1 (p.u.), indicating that the voltage levels are maintained within acceptable limits. In contrast, some other cases show slight deviations or lower values at certain busses, suggesting potential issues with voltage regulation or stability. Therefore, Case 4 can be considered favorable in terms of maintaining a stable and reliable voltage profile across the specified busses.

7.2.3 System cost (SC)

In this section, we discussed the progressive decrease in system cost from Cases 1 to 6 in comparison to the Base case, as depicted in **Figure** 7.

Cases 3, 4, and 6 stand out as they have the lowest system costs among all cases, with Case 6 achieving the highest reduction percentage at 98.97%. This indicates that Case 6 has the most optimized system formation in terms of cost savings. On the other hand, Cases 1, 2, and 5 also show significant cost reductions but are slightly higher compared to Cases 3, 4, and 6. Overall, **Figure** 7 shows that optimization strategies implemented in Cases 3, 4, and 6 have effectively reduced system costs while maintaining a high percentage reduction.

7.3 DE method results

7.3.1 Power loss reduction (PLR)

Figure 8 illustrates substantial decreases in active power losses across all cases when compared to the Base case, showcasing a range of improvements in different levels. Case 1 shows a substantial reduction in active power losses, achieving a percentage reduction of 98.62%. This reduction indicates an effective optimization strategy. Case 2 also demonstrates a noteworthy decrease in losses with a reduction percentage of 98.83%, indicating further improvement over Case 1. Case 3 (98.69%) exhibits a slightly lower reduction percentage compared to Case 2, but it still signifies a considerable enhancement in system performance. Case 4 and Case 5 both maintain a reduction percentage of 98.83% and 98.76%, showcasing consistent performance in reducing active power losses. While, Case 6 maintains the same reduction percentage

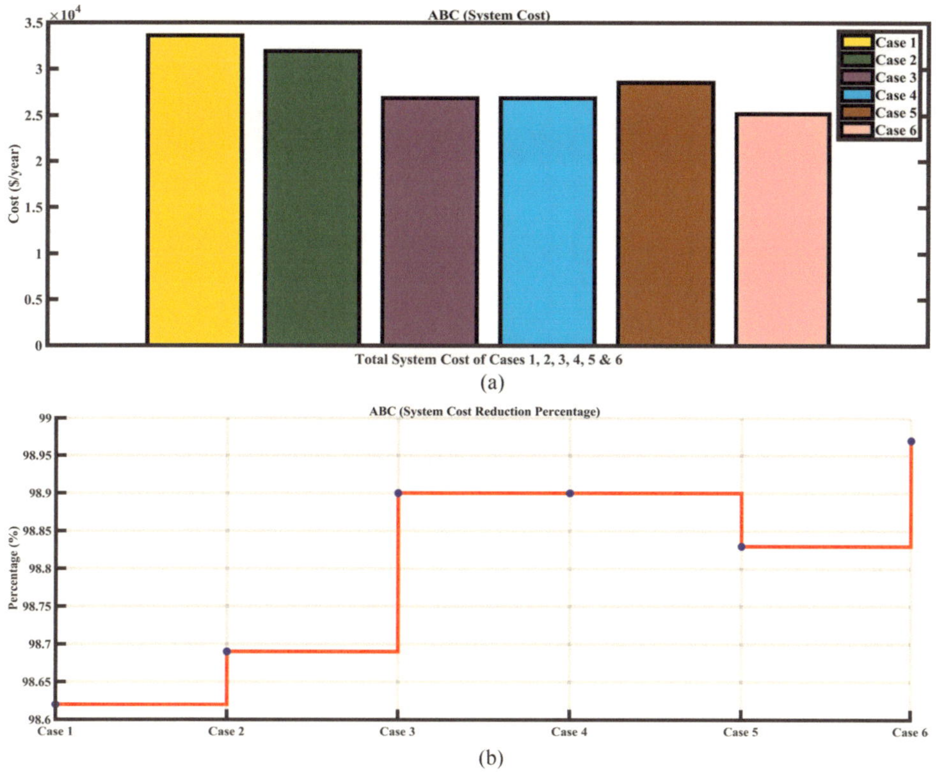

Figure 7.
Total system cost (a) and percentage reduction of system cost (b) of the ABC cases.

as Cases 2 and 4, which shows the lowest active power losses among all cases, indicating a highly optimized system configuration.

Overall, Cases 2, 4, and 6 demonstrate effective optimization strategies with notable reductions in active power losses, potentially leading to improved system efficiency and cost savings.

7.3.2 Voltage profile improvement (VPI)

Figure 9 depicts the assessment of voltage profiles across different cases in the power system. The cases, such as Cases 1, 2, 4, 5, and 6, exhibited consistent voltage levels across multiple busses when compared to the Base case.

Case 3 exhibited variances with increased voltages at specific busses, emphasizing its superior performance compared to all other cases. Overall, the assessment implies that the optimization strategies employed in Cases 1, 2, 4, 5, and 6 effectively handled voltage profiles.

7.3.3 System cost (SC)

Figure 10 shows the outcomes of DE cases. The initial system cost is $2,441,040 per year, serving as the baseline for comparative analysis. Case 1 shows an impressive system cost reduction to $33,600 per year, marking a remarkable 98.62% reduction.

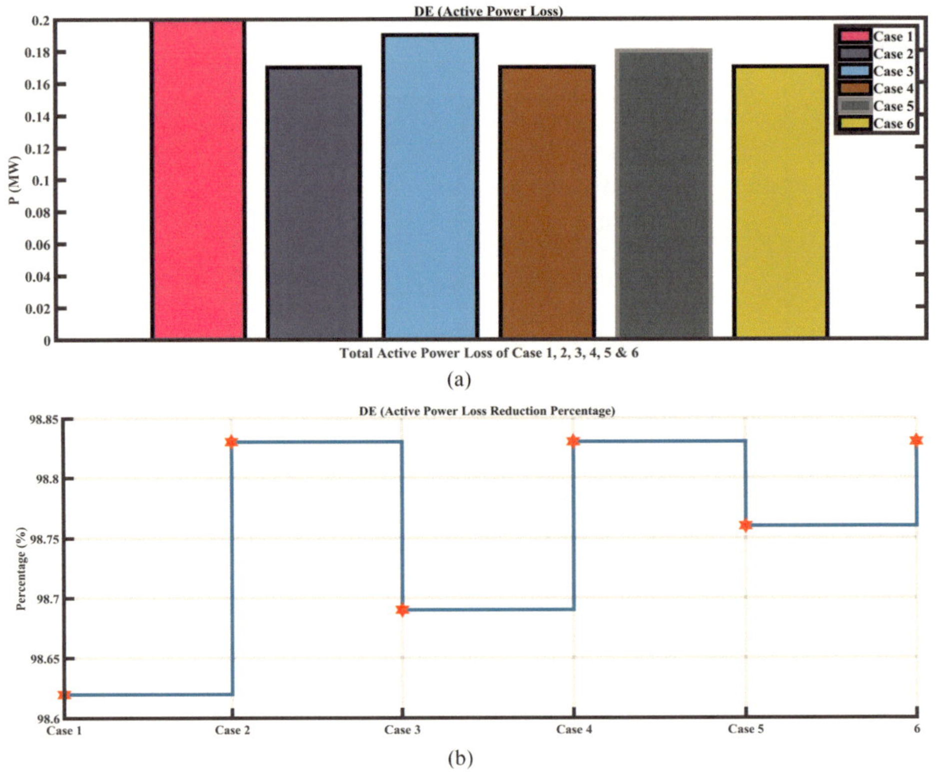

Figure 8.
Total active loss (a) and percentage reduction of active loss (b) of the DE cases.

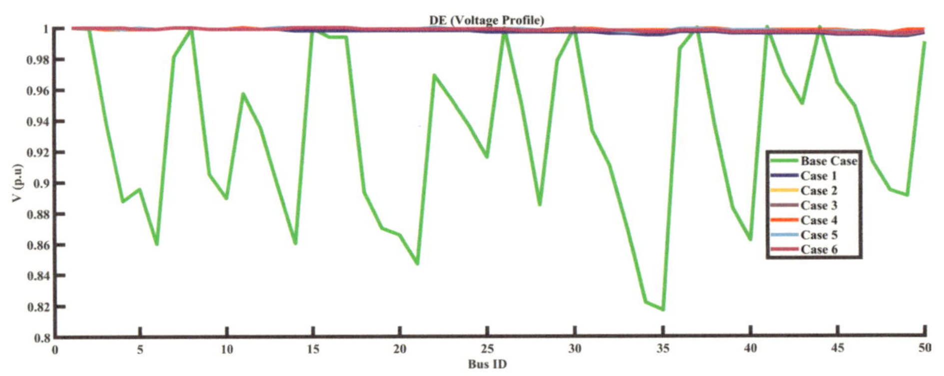

Figure 9.
Voltage profile improvement of all DE cases.

This outcome underscores the efficacy of the optimization strategy employed. Following closely, Case 2 achieves a substantial cost reduction to $28,560 per year, boasting a reduction percentage of 98.83% and presenting itself as marginally more cost-effective than Case 1.

In contrast, Case 3 reveals a system cost of $31,920 per year, with a commendable reduction percentage of 98.69%. Case 4 also demonstrates a significant cost reduction to $28,560 per year, aligning with Case 2 in its 98.83% reduction percentage.

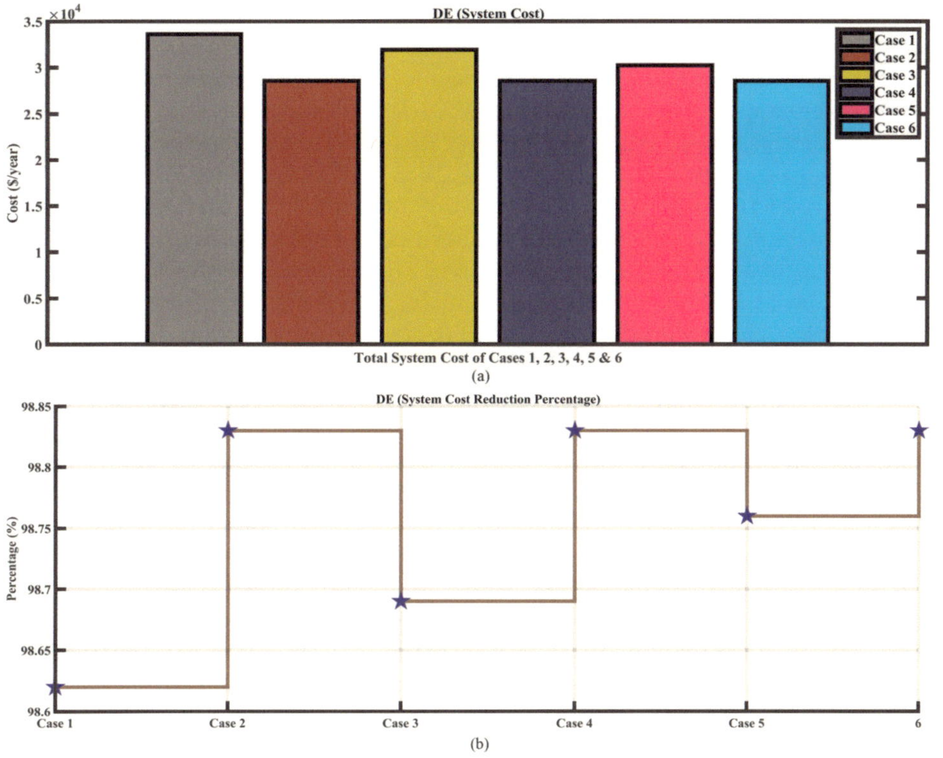

Figure 10.
Total system cost (a) and percentage reduction of system cost (b) of the DE cases.

Furthermore, Case 5 exhibits a system cost of $30,240 per year, achieving a reduction percentage of 98.76%, while Case 6 with a system cost of $28,560/year and a reduction percentage of 98.83%.

Notably, Cases 2, 4, and 6 emerge as standout performers, showcasing the highest cost reduction percentages while maintaining relatively lower system costs. However, a comprehensive evaluation of each case is crucial to determine the most suitable optimization strategy for the power system.

7.4 PSO, ABC, and DE method results

This section assesses the effectiveness of three metaheuristic techniques (DE, ABC, and PSO) on a 33 kV distribution system by analyzing power loss, voltage profiles, and annual system costs before and after the integration of RET. The Base case serves as the reference, while Cases 1 to 6 represent interventions in the 33 kV distribution system. The integration of RET in the Ashanti Region's 33 kV distribution system has reduced power losses and voltage drops, leading to lower annual system costs. The data presented in **Table 1** is related to the performance measurement of the 33 kV distribution system before and after the integration of RET. The performance parameters include the minimum and maximum bus voltages, total active power loss, RET location and power injection, system cost, and net savings.

Before integrating RET (Base case), the minimum and maximum bus voltages were 0.817 p.u. and 1 p.u., respectively. However, after the integration of RET, the minimum and maximum bus voltages improved significantly to 0.996 p.u. and 1 p.u., respectively, across all techniques (PSO, ABC, and DE). This improvement in voltage profile indicates that RET plays a significant role in maintaining the voltage magnitude within the international standard voltage range [51, 52], which is essential for a reliable power supply.

The total active power loss was also reduced significantly after the integration of RET. In PSO, this reduction ranged from 98.28 to 98.83% compared to the Base case. In ABC, the reduction fell within the range of 98.62 to 98.97%, and in DE, it ranged from 98.62 to 98.83%, all compared to the Base case. **Table 1** shows that the RET was installed on different busses in different cases. The results show that the location and power injection of RET have a significant impact on the system performance.

Furthermore, the integration of RET led to significant reductions in system costs. PSO showed annual system cost reductions ranging from 98.28 to 98.83%, ABC from 98.62 to 98.97%, and DE from 98.62 to 98.83%, demonstrating substantial cost savings. Notably, Case 6 in the ABC technique exhibited a more pronounced influence than all cases in PSO and DE.

These findings have substantial practical implications for real-world applications. The significant improvement in voltage profiles with RET integration underscores the potential for enhancing grid reliability and stability. Utility companies can leverage RET to ensure a stable voltage supply, adhering to international standards, thereby reducing the risk of power outages and improving service quality for consumers.

The considerable reduction in system costs through the use of optimization techniques like ABC, PSO, and DE highlights the economic benefits of RET. Policymakers and energy planners can adopt these techniques to minimize operational costs, making RET integration more financially viable and attractive.

The results demonstrate the importance of strategic placement and power injection of RETs to maximize system performance. Energy planners can use these insights to optimize the deployment of RET, ensuring that they are located where they can provide the most benefit in terms of voltage improvement and loss reduction.

By integrating RETs effectively, the energy sector can significantly reduce carbon emissions and contribute to environmental sustainability [6, 53]. This aligns with global efforts to combat climate change and promotes the transition to cleaner energy sources.

The methodologies and findings can be applied to other regions and distribution systems, especially in developing countries. This scalability ensures that the benefits of RET integration and optimization can be extended globally, aiding in the proper planning and optimization of distribution networks in various contexts.

By considering these adaptations and integrations, the research not only demonstrates the effectiveness of PSO, ABC, and DE but also paves the way for future innovations in optimizing power distribution systems, thereby making the content more significant and valuable for both researchers and practitioners in the field.

8. Conclusion

The integration of RET into the power system leads to a reduction in total active losses, cost savings, and a voltage profile that is substantially improved and in line with the IEC 60038-2009 voltage standard level [51, 52]. The performance improvement is highly dependent on the location and power injection of RET. Therefore, the results obtained in this chapter provide evidence of the efficiency and effectiveness of integrating RET into the distribution system in the Ashanti Region. The results obtained from this study offer valuable insights for decision-making regarding the possible implementation of RET in similar distribution systems. These findings can aid in optimizing system operations and attaining cost reductions.

Acknowledgements

We extend our heartfelt gratitude to God for guidance and blessings during this chapter's completion. Special thanks are due to Ing. Appiah Alexander Winston and Ing. Ansah Antwi Boasiako from ECG for providing data and to all lab technicians at AAMUSTED.

Conflict of interest

There is no conflict of interest in this study.

Notes/Thanks/Other declarations

Except for acknowledged quotations and references from published works, this chapter is entirely our original work and has not been submitted elsewhere in part or in full.

Author details

Prince Asabere[1*], Isaac Prempeh[1], Francois Sekyere[1], Willie K. Ofosu[2] and Patrick N. Ayambire[1]

1 Department of Electrical and Electronic Engineering, Ho Technical University, Ho, Ghana

2 School of Engineering Design, Technology, and Professional Programs, Penn State, Wilkes-Barre Lehman, PA, USA

*Address all correspondence to: princeasabere@yahoo.com

References

[1] Kabeyi MJB, Olanrewaju OA. Sustainable energy transition for renewable and low carbon grid electricity generation and supply. Frontiers Media S.A. 2022;**9**:1-19. DOI: 10.3389/fenrg.2021.743114

[2] Kober T, Schiffer HW, Densing M, Panos E. Global energy perspectives to 2060—WEC's world energy scenarios 2019. Energy Strategy Reviews. 2020;**31**: 100523. DOI: 10.1016/J.ESR.2020.100523

[3] Qazi A et al. Towards sustainable energy: A systematic review of renewable energy sources, technologies, and public opinions. IEEE Access. 2019; **7**:63837-63851. DOI: 10.1109/ ACCESS.2019.2906402

[4] Hansen K, Breyer C, Lund H. Status and perspectives on 100% renewable energy systems. Energy. 2019;**175**: 471-480. DOI: 10.1016/J. ENERGY.2019.03.092

[5] Ringkjøb HK, Haugan PM, Solbrekke IM. A review of modelling tools for energy and electricity systems with large shares of variable renewables. Renewable and Sustainable Energy Reviews. 2018;**96**:440-459. DOI: 10.1016/J.RSER.2018.08.002

[6] Al-Shetwi AQ. Sustainable development of renewable energy integrated power sector: Trends, environmental impacts, and recent challenges. Science of the Total Environment. 2022;**822**:153645. DOI: 10.1016/J.SCITOTENV. 2022.153645

[7] Abban J, Awopone A. Techno-economic and environmental analysis of energy scenarios in Ghana. Smart Grid and Renewable Energy. 2021;**12**(06): 81-98. DOI: 10.4236/sgre.2021.126006

[8] REN21. 2019. Renewables 2019 Global Status Report (Paris: REN21 Secretariat). ISBN 978-3-9818911-7-1. REN21. 2019. Renewables 2019 Global Status Report. [Online]. Available from: https://www.ren21.net/wp-content/uploads/2019/05/gsr_2019_full_report_en.pdf [Accessed: May 16, 2024]

[9] Choudhury S. Review of energy storage system technologies integration to microgrid: Types, control strategies, issues, and future prospects. Journal of Energy Storage. 2022;**48**:103966. DOI: 10.1016/J.EST.2022.103966

[10] Yoldaş Y, Önen A, Muyeen SM, Vasilakos AV, Alan İ. Enhancing smart grid with microgrids: Challenges and opportunities. Renewable and Sustainable Energy Reviews. 2017;**72**:205-214. DOI: 10.1016/J.RSER.2017.01.064

[11] Uchendu M. Placement of distributed generation and shunt capacitor in distribution network using cuckoo search algorithm. Nigerian Journal of Technological Development. 2020;**17**(2): 79-87. DOI: 10.4314/njtd.v17i2.2

[12] Nguyen TT, Truong AV, Phung TA. A novel method based on adaptive cuckoo search for optimal network reconfiguration and distributed generation allocation in distribution network. International Journal of Electrical Power & Energy Systems. 2016; **78**:801-815. DOI: 10.1016/J. IJEPES.2015.12.030

[13] Ogunsina AA, Petinrin MO, Petinrin OO, Offornedo EN, Petinrin JO, Asaolu GO. Optimal distributed generation location and sizing for loss minimization and voltage profile optimization using ant colony algorithm. SN Applied Sciences. 2021;**3**(2):3-8. DOI: 10.1007/s42452-021-04226-y

[14] Sultana U, Khairuddin AB, Mokhtar AS, Zareen N, Sultana B. Grey wolf optimizer based placement and sizing of multiple distributed generation in the distribution system. Energy. 2016; **111**:525-536. DOI: 10.1016/J.ENERGY. 2016.05.128

[15] Abdallah EM, Elsayed MI, ELgazzer MM, Hassan AA. Coyote multi-objective optimization algorithm for optimal location and sizing of renewable distributed generators. International Journal of Electrical and Computer Engineering. 2021;**11**(2):975-983. DOI: 10.11591/ijece.v11i2.pp975-983

[16] Pham TD, Nguyen TT, Dinh BH. Find optimal capacity and location of distributed generation units in radial distribution networks by using enhanced coyote optimization algorithm. Neural Computing & Applications. 2021;**33**(9): 4343-4371. DOI: 10.1007/s00521-020-05239-1

[17] PRP D, RVC V, Gowri MT. Ant lion optimization algorithm for optimal sizing of renewable energy resources for loss reduction in distribution systems. Journal of Electrical Systems and Information Technology. 2018;**5**(3): 663-680. DOI: 10.1016/J. JESIT.2017.06.001

[18] Pesaran HA, Nazari-Heris M, Mohammadi-Ivatloo B, Seyedi H. A hybrid genetic particle swarm optimization for distributed generation allocation in power distribution networks. Energy. 2020;**209**:118218. DOI: 10.1016/J.ENERGY.2020.118218

[19] Hadidian-Moghaddam MJ, Arabi-Nowdeh S, Bigdeli M, Azizian D. A multi-objective optimal sizing and siting of distributed generation using ant lion optimization technique. Ain Shams Engineering Journal. 2018;**9**(4):2101-2109. DOI: 10.1016/J.ASEJ.2017.03.001

[20] Institute of Electrical and Electronics Engineers, Proceedings 2012 2nd National Conference on Computational Intelligence and Signal Processing (CISP): 2-3 March 2012. IEEE, 2012

[21] Ramli MAM, Bouchekara HREH, Alghamdi AS. Optimal sizing of PV/ wind/diesel hybrid microgrid system using multi-objective self-adaptive differential evolution algorithm. Renewable Energy. 2018;**121**:400-411. DOI: 10.1016/J.RENENE.2018.01.058

[22] Kayalvizhi S, Vinod VK. Optimal planning of active distribution networks with hybrid distributed energy resources using grid-based multi-objective harmony search algorithm. Applied Soft Computing. 2018;**67**:387-398. DOI: 10.1016/J.ASOC.2018.03.009

[23] Kefayat M, Lashkar Ara A, Nabavi Niaki SA. A hybrid of ant colony optimization and artificial bee colony algorithm for probabilistic optimal placement and sizing of distributed energy resources. Energy Conversion and Management. 2015;**92**:149-161. DOI: 10.1016/J.ENCONMAN.2014.12.037

[24] Wright C, Sathre R, Buluswar S. The Global Challenge of Clean Cooking Systems. Dordrecht, Netherlands: Springer Science and Business Media B.V.; 2020. DOI: 10.1007/s12571-020-01061-8

[25] Keles S, Bilgen S, Kaygusuz K. Biomass Energy Source in Developing Countries. 2017. [Online]. Available from: www.journaleras.com

[26] I. Renewable Energy Agency, K. Development Bank, and D. Gesellschaft für Internationale Zusammenarbeit. The Renewable Energy Transition in Africa Powering Access, Resilience and Prosperity On behalf of the

[27] Scaling up Renewable Energy Deployment in Africa Detailed Overview

of Irena's Engagement and Impact. 2020. [Online]. Available from: www.irena.org

[28] G. Electric Company. 2022 Sustainability report a transformative era of action. 2022. [Online]. Available from: https://www.ge.com/investor

[29] Momodu AS. Energy use: Electricity system in West Africa and climate change impact. International Journal of Sustainable Energy Planning and Management. 2017;**14**:21-38. DOI: 10.5278/ijsepm.2017.14.3

[30] Monyei CG et al. Regional cooperation for mitigating energy poverty in sub-Saharan Africa: A context-based approach through the tripartite lenses of access, sufficiency, and mobility. Renewable and Sustainable Energy Reviews. 2022;**159**:112209. DOI: 10.1016/J.RSER.2022.112209

[31] Gibson Harold D. Taming a Wicked Problem: Energy Access Planning From an Energy-Poor Perspective. Cambridge, MA: Harvard Extension School, Harvard University; Master's thesis. 2017

[32] Asabere P, Sekyere F, Ayambire P, Ofosu WK. Optimal capacitor bank placement and sizing using particle swarm optimization for power loss minimization in distribution network. Journal of Engineering Research. 2024:4-5. DOI: 10.1016/J.JER.2024.03.007

[33] Veerasamy V et al. Recurrent network based power flow solution for voltage stability assessment and improvement with distributed energy sources. Applied Energy. 2021;**302**: 117524. DOI: 10.1016/J.APENERGY.2021.117524

[34] Grisales-Noreña LF, Garzon-Rivera OD, Ramírez-Vanegas CA, Montoya OD, Ramos-Paja CA. Application of the backward/forward sweep method for solving the power flow problem in DC networks with radial structure. Journal of Physics: Conference Series, Institute of Physics Publishing. 2020;**1448**:2-3. DOI: 10.1088/1742-6596/1448/1/012012

[35] Fonseca AG, Tortelli OL, Lourenço EM. Extended fast decoupled power flow for reconfiguration networks in distribution systems. IET Generation, Transmission and Distribution. 2018; **12**(22):6033-6040. DOI: 10.1049/iet-gtd.2018.5886

[36] SSN College of Engineering. Department of Electrical and Electronics Engineering, IEEE Power Electronics Society, and Institute of Electrical and Electronics Engineers, International Conference on Power and Embedded Drive Control: 16-18 March 2017, Chennai, India

[37] Pesaran M, Behnam H, MI. Studies in Systems, Decision and Control 262 Optimization of Power System Problems Methods, Algorithms and MATLAB Codes. [Online]. Available from: http://www.springer.com/series/13304

[38] Montoya OD, Gil-González W, Giral DA. On the matricial formulation of iterative sweep power flow for radial and meshed distribution networks with guarantee of convergence. Applied Sciences (Switzerland). 2020;**10**(17):4-10. DOI: 10.3390/app10175802

[39] Heidarifar M, Andrianesis P, Caramanis M. A Riemannian optimization approach to the radial distribution network load flow problem. Automatica. 2021;**129**:109620. DOI: 10.1016/J.AUTOMATICA.2021.109620

[40] Ghatak U, Mukherjee V. A fast and efficient load flow technique for

unbalanced distribution system. International Journal of Electrical Power & Energy Systems. 2017;**84**:99-110. DOI: 10.1016/J.IJEPES.2016.05.002

[41] Gad AG. Particle swarm optimization algorithm and its applications: A systematic review. Archives of Computational Methods in Engineering. 2022;**29**(5):2531-2561. DOI: 10.1007/s11831-021-09694-4

[42] Zhang X, Zou D, Shen X. A novel simple particle swarm optimization algorithm for global optimization. Mathematics. 2018;**6**(12):2-11. DOI: 10.3390/math6120287

[43] Jordehi AR. Enhanced leader PSO (ELPSO): A new PSO variant for solving global optimisation problems. Applied Soft Computing. 2015;**26**:401-417. DOI: 10.1016/J.ASOC.2014.10.026

[44] Nagra AA, Han F, Ling QH. An improved hybrid self-inertia weight adaptive particle swarm optimization algorithm with local search. Engineering Optimization. 2019;**51**(7):1115-1132. DOI: 10.1080/0305215X.2018. 1525709

[45] Kiani AT, Nadeem MF, Ahmed A, Khan IA, Alkhammash HI, Sajjad IA, Hussain B. An Improved Particle Swarm Optimization with Chaotic Inertia Weight and Acceleration Coefficients for Optimal Extraction of PV Models Parameters. Energies. 2021. DOI: 10.3390/en14112980

[46] Ghambari S, Rahati A. An improved artificial bee colony algorithm and its application to reliability optimization problems. Applied Soft Computing. 2018;**62**:736-767. DOI: 10.1016/J. ASOC.2017.10.040

[47] Garg H, Rani M, Sharma SP. An efficient two phase approach for solving reliability–redundancy allocation problem using artificial bee colony technique. Computers & Operations Research. 2013;**40**(12):2961-2969. DOI: 10.1016/J.COR.2013.07.014

[48] Al-Ammar EA et al. ABC algorithm based optimal sizing and placement of DGs in distribution networks considering multiple objectives. Ain Shams Engineering Journal. 2021;**12**(1):697-708. DOI: 10.1016/j.asej.2020.05.002

[49] Mohamed AW. A novel differential evolution algorithm for solving constrained engineering optimization problems. Journal of Intelligent Manufacturing. 2018;**29**(3):659-692. DOI: 10.1007/s10845-017-1294-6

[50] Liao TW. Two hybrid differential evolution algorithms for engineering design optimization. Applied Soft Computing. 2010;**10**(4):1188-1199. DOI: 10.1016/J.ASOC.2010.05.007

[51] Ackermann T, Knyazkin V. Interaction between distributed generation and the distribution network: Operation aspects. In: Proceedings of the IEEE Power Engineering Society Transmission and Distribution Conference, Vol. 2, no. ASIA PACIFIC. Piscataway, NJ: IEEE; 2002. pp. 1357-1362. DOI: 10.1109/tdc.2002.1177677

[52] Standard H, Horizontale N. International standard Internationale IEC standard voltages. 2009. [Online]. Available from: https://webstore.iec.ch/publication/72877

[53] Razmjoo A, Gakenia Kaigutha L, Vaziri Rad MA, Marzband M, Davarpanah A, Denai M. A technical analysis investigating energy sustainability utilizing reliable renewable energy sources to reduce CO_2 emissions in a high potential area. Renewable Energy. 2021;**164**:46-57. DOI: 10.1016/J. RENENE.2020.09.042

Chapter 2

Decreasing DG Generation Uncertainty Prediction in Energy System Management

Farshid Keynia and Golamreza Memarzadeh

Abstract

Renewable energy sources play an important role in most countries of the world as an important part of electric energy production. Today, wind and solar energy is growing and expanding dramatically in the world. These two energy sources are among the most economically attractive renewable sources that are available for electricity production today. Integrating substantial wind and solar energy into a power system presents significant operational challenges, leading to fluctuations in electricity prices. The primary challenge with wind energy, compared to traditional energy sources, is its high uncertainty in production. This point has a significant effect on the profitability of these sources of energy production in the electricity market. The time series of renewable resources usually has a complex behavior. Among the complexities of its graph is its non-linear and variable behavior. Therefore, based on this unstable, intermittent, and uncertain behavior, an accurate and robust prediction model should be designed and implemented. In other words, it should be able to overcome these uncertainties with various methods. Among the suitable ways to have high accuracy to predict the parameters of wind production and the price of electric energy is the use of accurate and combined methods for forecasting. Particularly, the parameter of wind production, which in long-term horizons, the error of its prediction with simple methods will be very high. Among these accurate methods, we can mention prediction based on deep learning models and interval prediction.

Keywords: decentralized generation, wind and solar energy generation, neural networks, machine learning, energy generation prediction, smart grids

1. Introduction

As it was said, the uncertainty of renewable resources has a significant impact on the performance of the producer in the power system, especially in the electricity market but increasing renewable energy generations are very important to compete with traditional energy sources [1].

Due to their intermittent and random fluctuations, it will be challenging to use and exploit their power [2, 3]. The accurate prediction of the production of these resources in the short term plays an important role in solving the challenges. However, the non-linear and non-stationary characteristics of the time series behavior of these sources make it difficult to accurately model and predict. For instance, the wind power generated by a wind farm heavily relies on the unpredictable nature of wind speed. Sudden fluctuations in wind power output can lead to increased operating costs for the power system. Moreover, the uncertainty of wind speed makes it difficult to predict how much power will be produced. Thus, accurate wind speed predicting can reduce the costs of wind power producers and also helps them to be more effective in the electricity market. Recently, many investigators have proposed different approaches for wind speed prediction. In some of these methods, they employed meteorological and geographical data to predict wind speed, but they are not accurate enough in short-term wind speed prediction [3, 4]. Some of the methods that are used in short-term wind speed prediction are tuned to find a relationship between the input and output variables. These approaches are used for the statistical methods [5–7]. These models have more accurate results than physical model in wind speed prediction. These methods are more applicable because they are capable of understanding the physical properties of the mechanisms. But they are not capable of modeling non-linear behaviors of wind speed series. The third approach is spatial correlation methods, which can achieve greater prediction accuracy in some conditions [8, 9]. However, artificial intelligence-based approaches such as artificial neural networks (ANN) [10–13], support vector machine (SVM) [14], and fuzzy logic [15] are more accurate in short-term wind speed prediction. Among neural networks, deep-learned neural networks are much more efficient in short-term wind speed prediction [16–19]. Also, these neural networks are remarkably accurate in prediction various parameters, such as electrical load and price [20, 21].

In addition, the uncertainty and intermittency of wind speed has a great impact on wind power generation, which can severely negatively affect the profitability of its presence in the electricity market. Therefore, accurate prediction of wind speed is very important for wind farm operation and planning. However, electrical energy differs from other commodities in several key ways. It cannot be stored in large quantities, and maintaining the stability of the power system requires a constant balance between generation and consumption. In short timeframes, most electricity users are unaware of or indifferent to its price. Transmission bottlenecks often restrict the flow of electricity between regions, which can lead to price fluctuations or even increases in the electricity market. Additionally, factors such as fuel price volatility, load uncertainty, variations in hydropower production, production uncertainty (e.g., outages), and the behavior of market participants contribute to electricity price uncertainty [22–24]. Therefore, price forecasting is crucial in today's electricity markets. Companies that trade in these markets rely heavily on price forecasting techniques to place bids or hedge against volatility. Therefore, in order to model the uncertainty of the presence of a renewable producer, one of the most important solutions is to rely on the accurate estimation of the critical parameters of the system, that is, their production and the price of electric energy [25–27].

2. Regulatory and policy implications

The use of renewable energy on the indicators of employment, gross domestic product, added value, and economic well-being has a key role. The results of the studies indicate that the development of renewable energy has created more than

1.8 million jobs in the world, and in 2015, the share of Asian countries is more than 60%. Although the development of renewable energy has caused an average increase of 0.63% in the growth of the world's GDP, the countries with oil reserves have seen negative growth. The added value obtained on average is 610 dollars for each kilowatt of module production in the solar energy sector and 583 dollars for each kilowatt in the wind energy generation equipment manufacturing sector. The impact of renewable energy development on economic well-being in different countries has been positive. The energy production from renewable sources, in addition to meeting energy demand and replacing them with part of the demand for fossil fuels, brings positive short-term and long-term economic effects with different shares. Therefore, the enhancement of macroeconomic indicators is seen as one of the driving forces behind the development of policies aimed at increasing energy production.

3. Impact on energy markets

Many forecasting methods have been proposed recently as important functions in the deregulated electricity power markets. The forecasted values of effective variables such as electricity load, renewable power generations and its prices have a key role in vital decision making and scheduling for power market participants. On the other hand, based on volatile, intermittent, and uncertain behavior of these variables, accurate, and robust forecast model should be designed.

4. Technological limitations and advancements

As mentioned before, the uncertainty of renewable resources and the price of electric energy have a significant impact on the performance of these producers in the electricity market. Due to intermittent and random fluctuations of wind and sun, their use and exploitation will be challenging. Accurate forecasting of renewable production in the short term plays an important role in solving challenges. Accurate prediction of these resources can effectively reduce the risk of renewable energy failure in hybrid power systems. However, the non-linear and non-stationary characteristics of renewable resources make it difficult to accurately model and predict. For example, the wind power produced by a wind farm is highly dependent on the random nature of wind speed, and unexpected changes in wind power output increase the operating cost of the power system. In addition, the uncertainty and intermittency of wind speed has a great impact on wind power generation, which can severely negatively affect the profitability of its presence in the electricity market. Therefore, accurate prediction of wind speed is very important for wind farm operation and planning. Regarding the presence of wind generators in the electricity market, wind generation is not the only critical parameter of the system. Another of these parameters is the price of electric energy. However, electricity differs significantly from other commodities. It cannot be stored substantially, and maintaining power system stability requires a constant balance between generation and load. In short timeframes, most electricity users are unaware of or indifferent to its price. Transmission bottlenecks often limit the transportation of electricity between regions, leading to price fluctuations or even increases in the electricity market. Additionally, factors such as fuel price fluctuations, load uncertainty, hydropower production variability, production uncertainties (such as outages), and market participants' behavior contribute to electricity price uncertainty. Price forecasting is crucial in today's electricity markets. Companies trading in these markets use price forecasting techniques extensively to bid or hedge against volatility. Accurately estimating critical parameters, such as renewable power production and electricity prices, is essential for modeling the uncertainty of renewable energy producers.

Due to the high uncertainty of renewable sources and on the other hand, their uncertainty and probability, forecasting and planning wind production, especially for long horizons, are a difficult and complicated matter. Therefore, accurate estimation of wind speed and production can make their presence in the electricity market more profitable. The negative environmental impact of fossil fuels has made many countries in the world interested in investing in renewable energy. On the other hand, governments have increased their support for replacing wind energy with fossil fuels. In general, low environmental pollution and low operating and maintenance costs are the reasons for increasing attention to renewable energies. On the other hand, the variable nature of renewable resources makes them difficult to predict. In recent years, the strategic presence of wind producers in the electricity market has been the focus of researchers in many studies. Restructured electricity markets were originally designed for traditional electricity producers, where power plants offer their output to the electricity market a few hours before actual delivery. Then, the independent operator of the system determines the market price and production share of each power plant. However, for wind producers, due to the uncertainty in their production, the proposed amount will be different from the production amount. Due to the significant interest in the effective and economical integration of renewable resources in the electricity market, many researchers in the energy field have focused on this area. Wind energy producers, in particular, aim to maximize their profits from the sale of generated electricity. However, the inherent uncertainties in wind energy production and market prices introduce complexities in strategically participating in the electricity market. So, in other words, due to the uncertainty of wind production, the unplanned presence of these power plants can sometimes cause losses to the producer. Therefore, according to the mentioned materials, determining how the renewable power plant is present in the electricity market to reach the maximum amount of profit requires the examination of important issues and points. There are various methods to reduce the uncertainty of wind production and its negative effects on the presence of wind power plants in the electricity market. Among these things, it is to predict with reasonable accuracy the production of these power plants and then provide this amount to the electricity market. Of course, due to the dynamic and variable nature of wind generation, predicting the amount of electrical energy produced is a complicated matter. Another way in this field is to produce a large number of scenarios for the production of wind electric energy and then reduce these scenarios to a smaller number and then the presence of renewable power plants in the electricity market according to these predetermined scenarios.

The uncertainty of wind energy production and electricity prices directly impacts the discrepancy between actual and predicted profits, ultimately reducing the power plant's profitability. To address this issue, accurate forecasts of wind power production and electricity prices for the following day are essential. By employing modern and robust forecasting methods, it becomes possible to predict wind power output and electricity prices with acceptable accuracy, thereby enabling the renewable power plant to achieve satisfactory profits. Deep neural networks are among the powerful techniques used for forecasting wind production, prices, and other crucial parameters of the power system. One of the types of deep neural network methods that have attracted a lot of attention for predicting time series today is the LSTM neural network. These neural networks have considerable ability and acceptable accuracy in predicting non-deterministic parameters. Meanwhile, by combining this network with pre-processing methods such as choosing the appropriate input and analyzing the signal, the prediction accuracy of these prediction engines can be significantly increased [22, 23].

Input selection is a common process in machine learning where a subset of the available data features is chosen for use by the learning algorithm. The main idea of the proposed feature selection method is to select a subset of entries by removing irrelevant and redundant entries. Among the methods of input selection, we can mention correlation analysis, numerical sensitivity analysis, and principal component analysis, which are all linear methods of input selection. One of the most advanced input selection techniques is the mutual enhancement method, which is based on the concept of information entropy.

Along with input selection, signal analysis will be one of the most important methods to improve prediction accuracy. Signal analysis is a key step in predicting the critical parameters of the energy system. The effective extraction of signal features is used to reveal the fundamental physical nature of a phenomenon, to fully understand the dynamical characteristics of a system, and to provide compelling evidence for predictive diagnosis. However, real signals are usually very complex, and this complexity contains rich information about the state and fundamental information of the signal being analyzed for prediction. Therefore, complex signal analysis is a key research topic and plays an important role in predicting critical parameters of energy system. To date, various signal analysis methods have been proposed for uncertainty prediction, including Fourier transform, wavelet, empirical signal decomposition, variable mode decomposition, and local average decomposition.

The combination of CNN and LSTM networks is a powerful synergy for effective management of time series data. While CNN layers excel at extracting valuable information and learning internal representations in time series data, LSTM layers effectively understand short-term and long-term dependencies. Accordingly, using the CNN-LSTM model, it will be useful and effective to enhance multi-stage renewable energy forecasting by taking advantage of the strengths of CNN and LSTM models. The first component of the hybrid prediction model is the CNN layer, where complex mathematical operations are applied to extract essential features from the input variables. This initial step plays an important role in identifying important patterns in the data. The next components focus on the features generated by the LSTM layers. These layers are a suitable tool for capturing time dependencies and information covering different time periods. This mechanism enables the model to focus on specific aspects of the data and prioritize the most relevant information for predictive accuracy [24–28].

5. Economic analysis

To investigate the impact of advanced forecasting methods on the performance of renewable resources in the electricity market, the performance of a wind power plant in the electricity market has been investigated according to different forecasting methods. In other words, the impact of forecasting with different accuracy on the profitable performance of wind power plant has been described and elaborated.

To achieve this, wind power production and electricity prices were predicted for a wind power plant with a specific capacity over a given period. The forecasted results for wind power production during the 4 months of 2018 are presented in **Table 1**.

The actual and forecasted wind production for these months of 2018 is shown in **Figure 1**. It is observed that the actual wind production follows the predicted values with acceptable accuracy.

Table 2 displays the MAPE error for predicting electricity prices during this month. As indicated, the MRMI-LSTM network outperforms the MRMI-MLP neural network in terms of MAPE value for electricity price prediction. **Figure 2** illustrates the predicted and actual electricity prices for these months in 2018.

Figure 3 illustrates both the anticipated and actual hourly profits derived from the predicted values of wind production and electricity prices. The figure reveals that the expected and actual profits align closely for most hours of the day. Additionally, **Table 3** details the forecasted profits generated by wind turbines in the electricity market during the final days of the designated months.

Forecasting method	Index	March	June	September	December
MRMI-MLP	MAPE	63.7725	64.5319	114.1421	71.3411
MRMI-GRU	MAPE	41.2349	45.5329	48.3872	36.7461
MRMI-LSTM	MAPE	36.6766	38.8922	41.5218	32.8559

Table 1.
Results of wind power production forecasting in PJM 2018 [22].

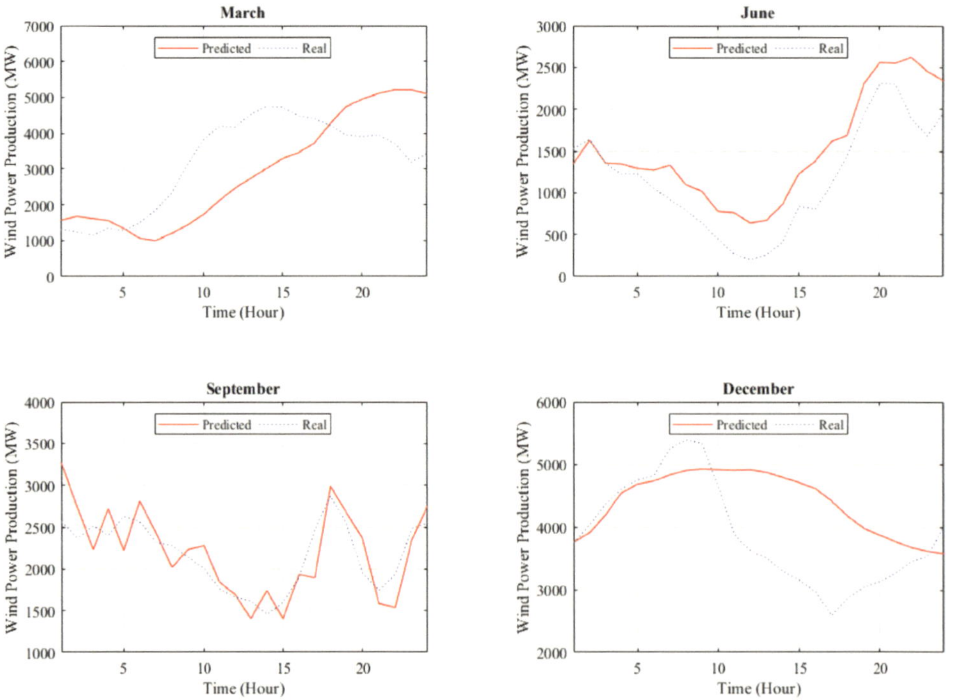

Figure 1.
Predicted and real wind power production of the year 2018.

Forecasting method	Index	March	June	September	December
MRMI-MLP	MAPE	51.8113	44.7816	43.356	28.3011
MRMI-GRU	MAPE	18.6157	20.4026	14.6383	9.7172
MRMI-LSTM	MAPE	17.7829	16.5913	13.5835	9.2351

Table 2.
Results of electricity price forecasting in PJM 2018 [22].

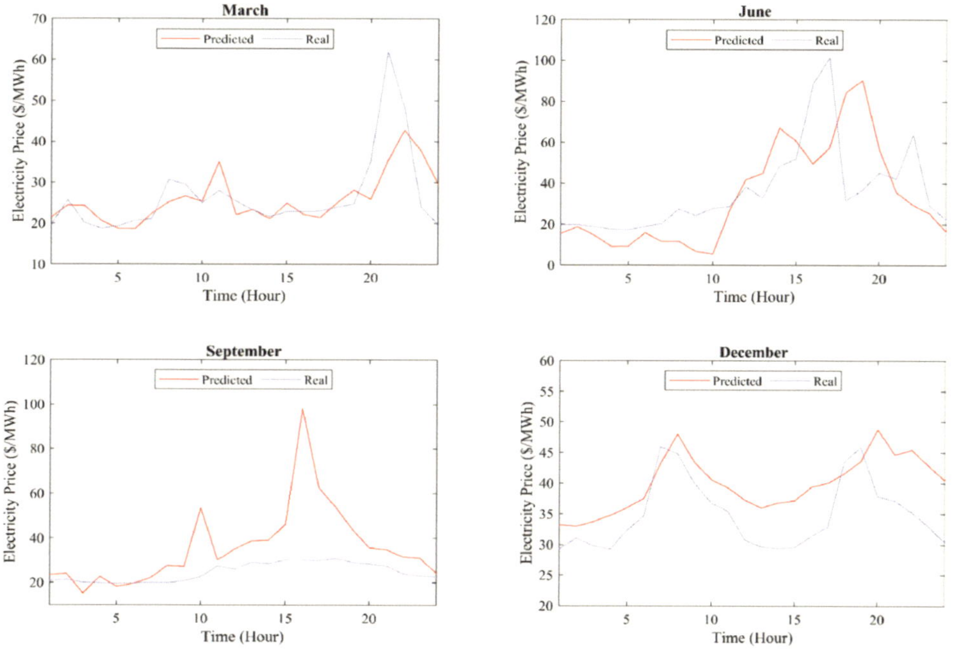

Figure 2.
Predicted and real electricity price of the year 2018.

Therefore, according to the results stated in **Tables 1** and **2**, we have realized that the use of different forecasting methods with different accuracy and performance has a direct impact on the profitability of these resources in the electricity market. In the same way, in other different structures of the power system, its prediction and accuracy have a direct impact on economic indicators. For example, in the discussion of electric load forecasting, with a 1% reduction in error, we will improve profitability for the power system by 0.1 to 0.4%. This will also affect the performance of renewable resources. But its quantity and quality will be different for different sources and different sectors.

6. Prediction-based modeling of uncertain renewable system parameters

The issue of the uncertainty of renewable energy and the price of electric energy directly affect the difference between the actual amount of profit and their predicted amount. This point will reduce the profit of the power plant. To overcome this problem, an accurate forecast of the amount of renewable production power for the

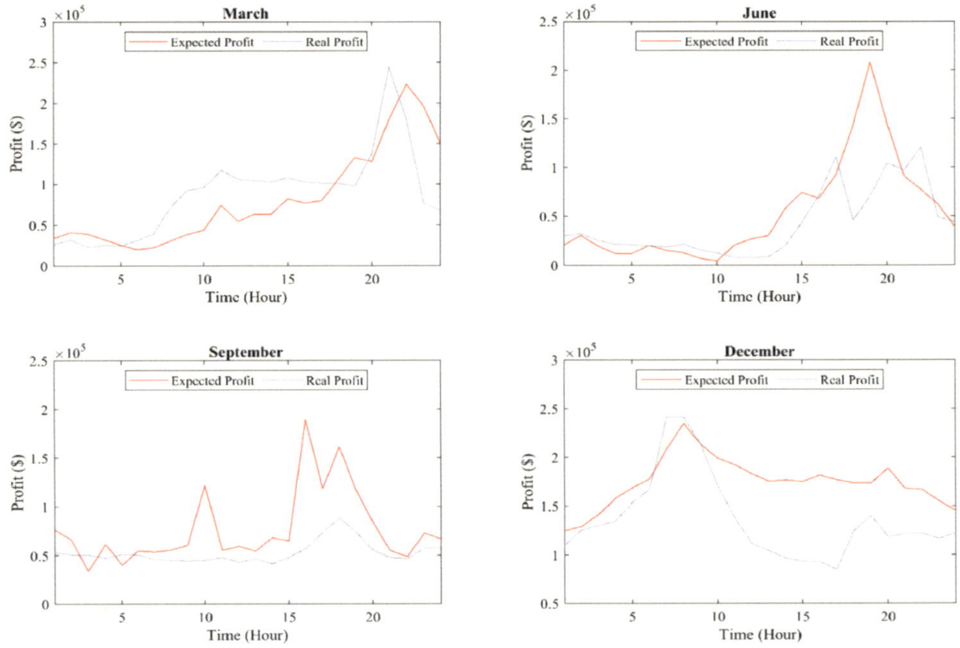

Figure 3.
Expected hourly and real profit.

Month	Real profit ($)	Originally expected profit ($)
March	2685788.51	3,014,495
June	1,454,469	1,010,498
September	455039.1	604088.2
December	1,498,061	1,757,419

Table 3.
Real profit, originally expected profit of the year 2018 [22].

next day will be helpful. Among the powerful methods for predicting renewable production and other important parameters of the power system are neural networks. Neural networks and especially deep neural networks have considerable ability and accuracy in predicting uncertain parameters. Meanwhile, by combining this network with pre-processing methods such as selecting the appropriate input and analyzing the signal, the prediction accuracy of these prediction engines can be significantly increased. In addition to these cases, the combination of neural networks with other forecasting methods in the form of a group learning method has a significant effect on improving the accuracy of the estimator. In other words, relying on this approach, various forecasting methods in a combined space can cover each other's weaknesses and improve forecasting accuracy in a tangible way.

6.1 Feature selection

Input selection is a typical process in machine learning, where a subset of the available data features is chosen for use by the learning algorithm. The main objective of the proposed feature selection method is to eliminate irrelevant and redundant inputs, retaining only the most pertinent features. Common methods for input selection include correlation analysis, numerical sensitivity analysis, and principal component analysis, all of which are linear techniques. One of the more advanced input selection techniques is the mutual information method, which relies on the concept of information entropy. Some of these methods used to identify the best inputs are described below.

6.1.1 MI-based feature selection

One of the most advanced input selection techniques is the mutual information method, which relies on the concept of information entropy. This section outlines the proposed approach for selecting the most relevant inputs, detailing the implementation of a two-stage input filter designed to identify the best features [25].

- Relevance filter: Evaluating the similarity between input and output data is essential for input selection. Mutual Information (MI) is an effective method for measuring this similarity. The MI between input and output is denoted by $I(x_i, y)$, which is based on the computation of the joint probability distribution of $P(x,y)$ and the independent probability distribution of $P(x)$ and $P(y)$. assuming that:

$$SC = \{x_1, x_2, x_3, \ldots x_M\} \quad (1)$$

is a set of inputs and y is the output to be predicted. The relationship between the output y and each of the x_i is called $D(x_i)$, and in other words, it expresses the degree of similarity of an input with the output of the problem, and its relationship is equal to:

$$D(x_i) = I(x_i, y) \quad (2)$$

Also, $I(x,y)$ based on mutual information between two random variables x and y is calculated through Eq. (3):

$$MI \quad (3)$$

- Redundancy filter: In most of the past researches, all common information between two inputs is proposed to calculate the similarity. It has been discussed in various references that when two inputs x_i and x_j are very similar, one of these two inputs can be selected. In other words, the two selected inputs may have a very high similarity to each other, while these two may have little similarity to the prediction output. In other words, the similarity between two data inputs is assessed, and if their similarity exceeds a certain threshold, removing one input can improve the algorithm's convergence without hindering the simulation process. However, if the threshold is set too high, few inputs may be identified for selection. The calculation of similarity or the same $\Re(x_i, x_j)$ is presented in the following form:

$$\Re(x_i, x_j) = I(x_i, x_j) \quad (4)$$

6.1.2 The effect of feature selection on prediction

As mentioned, input selection significantly increases the speed and accuracy of the predictor. To demonstrate this, a predictive model based on LSTM network is trained to predict a future step of wind speed. In two models, one of which includes the presence of input selection and the other without using this pre-processing method, the results are presented in **Table 4**. The results in this table demonstrate that the MI-LSTM model outperforms the LSTM method in short-term wind speed forecasting. This is evidenced by the lower values of MAPE, RMSE, and MAE in the MI-LSTM model. Essentially, incorporating input selection blocks into the LSTM network has improved both the accuracy and speed of this prediction method.

6.2 Signal decomposition

Signal analysis is a key step in predicting critical parameters of the energy system. Effectively extracting signal features helps to uncover the fundamental physical nature of a phenomenon, fully understand the dynamic characteristics of a system, and provide compelling evidence for predictive diagnosis. However, real signals are usually very complex, containing rich information about the state and essential details necessary for accurate prediction. Therefore, complex signal analysis is a vital research topic and plays an important role in predicting critical parameters of energy systems. To date, various signal analysis methods have been proposed for uncertainty prediction, including Fourier transform, wavelet, empirical signal decomposition, variable mode decomposition, and local average decomposition. In the following, some of the most important methods are examined.

6.2.1 Wavelet transform

Time series patterns in renewable energy are frequently non-linear and dynamic, showing sharp fluctuations that greatly affect predictions. Wind and electricity price signals feature variable characteristics, including level shifts, slopes, and seasonal trends, which add complexity to the analysis. To tackle these challenges, wavelet transforms are used to effectively examine the asymmetric nature of the data. Wavelet transform can extract different aspects of the data, such as breakpoints and discontinuities, making them a valuable tool for managing complex and variable signals. Usually, the discrete wavelet transform is presented to improve the calculations, and its relationship is as follows:

$$wavelet(v,k) = 2^{-\left(\frac{v}{2}\right)} \sum_{t=0}^{T-1} g(t)\varphi\left[\frac{\left(t - k * 2^{v}\right)}{2^{v}}\right] \tag{5}$$

Forecasting method	Number of the input feature	MAE (m/s)	RMSE (m/s)	MAPE (%)	Run time (sec)
LSTM	400	2.0018	2.6535	28.0287	1461.23
MI-LSTM	12	1.4308	1.8871	21.5643	686.12

Table 4.
The results of different forecasting methods.

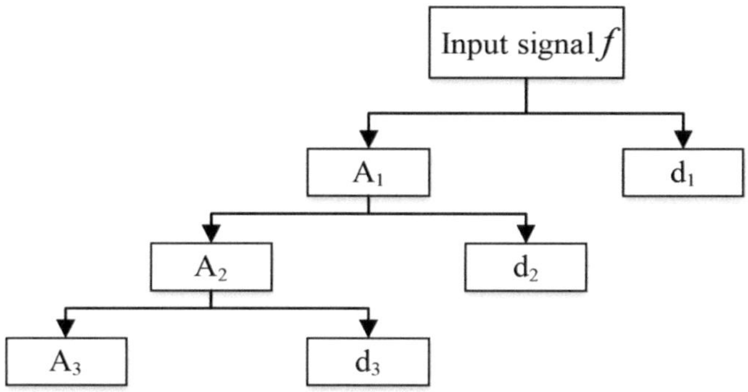

Figure 4.
Data separation using DWT.

In this equation, k is the transition variable, v is the scaling variable, $g(t)$ is the wave that must be discretized by the wavelet transform, $\varphi(\bullet)$ is the mother function of the wavelet transform, t is the discrete time interval, and T is the length of the signal to be decomposed. Among the most famous wavelets, we can mention Haar, Dabichiz, Morelet, Gussi, and Simlets. In the introduced Violet families, Dabichiz is more popular and efficient. Dobbychase type one is Violet Har. How to separate input data using wavelet transformation is shown in **Figure 4**. The discrete wavelet transforms returns two sets of coefficients as output, approximation coefficients, and detail coefficients. The approximation coefficients indicate the output of the low-pass filter (averaging filter) in the discrete wavelet transform, and the detail coefficients indicate the output of the high-pass filter (derivative filter) in the discrete wavelet transform. Following is the implementation of the wavelet transform for a given signal with the db4 wavelet function:

6.2.2 The effect of signal decomposition on prediction

Or as another example, by adding the wavelet transform block and selecting the input simultaneously to the LSTM network, the prediction accuracy has been significantly improved (**Table 5**).

These results are also confirmed in **Figure 5**, where the actual and predicted wind speed results with the above six methods are presented. As can be seen, the WT-FS-LSTM method followed the actual wind speed with extremely high accuracy.

7. Forecasting module

Various innovative methods have been presented for forecasting in modern electrical energy systems, and each of these methods has its own efficiency according

LSTM	128	2.0018	2.6535	28.0287
FS-LSTM	128	1.4308	1.8871	21.5643
WT-LSTM	128	0.6673	0.8739	10.0586
WT-FS-LSTM	128	0.1973	0.2700	3.1271

Table 5.
The results of different forecasting methods.

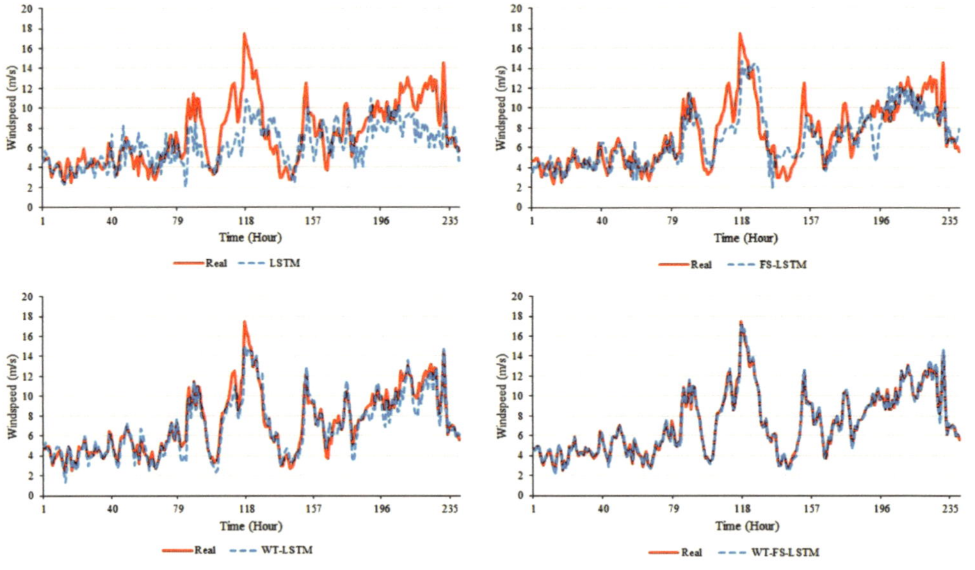

Figure 5.
The comparison of a different forecasting method.

to the type of data and forecasting horizon. These methods have a wide range and each of them has its own characteristics and structure. In this section, different neural networks have been investigated.

7.1 Neural networks

Over the past few decades, advancements in computing have facilitated the development of algorithms that simulate the computational behavior of the human brain. This progress has spurred extensive research among computer scientists, engineers, and mathematicians, leading to significant developments in artificial intelligence, particularly in computational intelligence, which includes artificial neural networks (ANNs). Within this field, various mathematical and software models inspired by the human brain have been created to tackle a wide range of scientific, engineering, and practical challenges across multiple disciplines. A neural network, as the name suggests, consists of interconnected nodes, each representing a processing unit. The connections between these nodes establish logical relationships. Overall, an artificial neural network functions as a parallel distributed processor made up of simple units that learn from empirical knowledge. By adjusting inter-unit connections, the network effectively builds and utilizes this knowledge to solve complex problems.

An artificial neuron is an information processing unit that is the basis of neural network operation. **Figure 6** shows the model of an artificial neuron. This neuron contains three main elements as follows:

1. A set of weight coefficients and biases, which are specified by Wi and b, and which are determined for each neuron and input associated with specific weight coefficients.

2. The adder is used to sum the input signals Xi which are weighted by Wi.

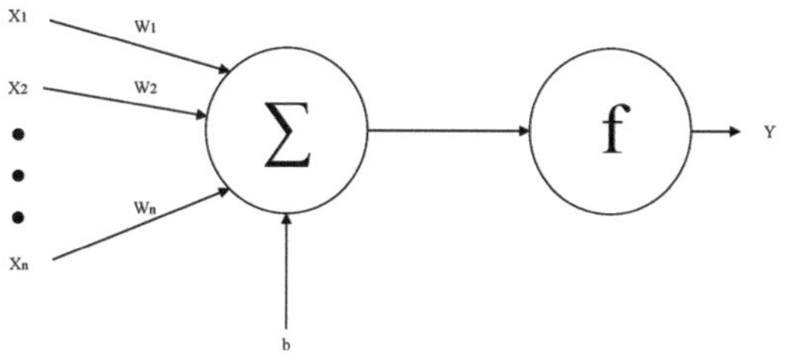

Figure 6.
An artificial neuron model.

3. The activation function f, which is used to limit the output range of a neuron. The activator function has various types, some of which are: symmetry hard limit, linear saturator, Sigmoid, Log-Sigmoid.

7.2 LSTM deep neural network

A Recurrent Neural Network (RNN) is a type of deep neural network. One major issue with RNNs is the vanishing gradient problem. To address this, the Long Short-Term Memory (LSTM) network was introduced. LSTM networks combine short-term and long-term memory capabilities, allowing them to handle both types of dependencies and mitigate the vanishing gradient problem. A key component of the LSTM network is the memory cell, which manages and retains information over extended sequences. The basic structure of an LSTM cell is illustrated in **Figure 7**.

According to **Figure 7**, the basic LSTM cell contains three gate units: the input gate, output gate, and forget gate. The processes for updating the cell state and calculating the LSTM outputs are illustrated below:

$$f_t = \sigma\left(W_{xf}.x_t + W_{hf}.h_{t-1} + b_f\right) \tag{6}$$

$$i_t = \sigma\left(W_{xi}.x_t + W_{hi}.h_{t-1} + b_i\right) \tag{7}$$

$$\overline{c_t} = tanh\left(W_{xc}.x_t + W_{hc}.h_{t-1} + b_c\right) \tag{8}$$

$$c_t = f_t^{*}c_{t-1} + i_t^{*}\overline{c}_t \tag{9}$$

$$o_t = \sigma\left(W_{xo}.x_t + W_{ho}.h_{t-1} + b_o\right) \tag{10}$$

$$h_t = o_t^{*}tanh\left(c_t\right) \tag{11}$$

$$y_t = \sigma\left(W_{hy}.h_t + b_y\right) \tag{12}$$

$$\sigma(x) = \frac{1}{1+e^{-x}} \tag{13}$$

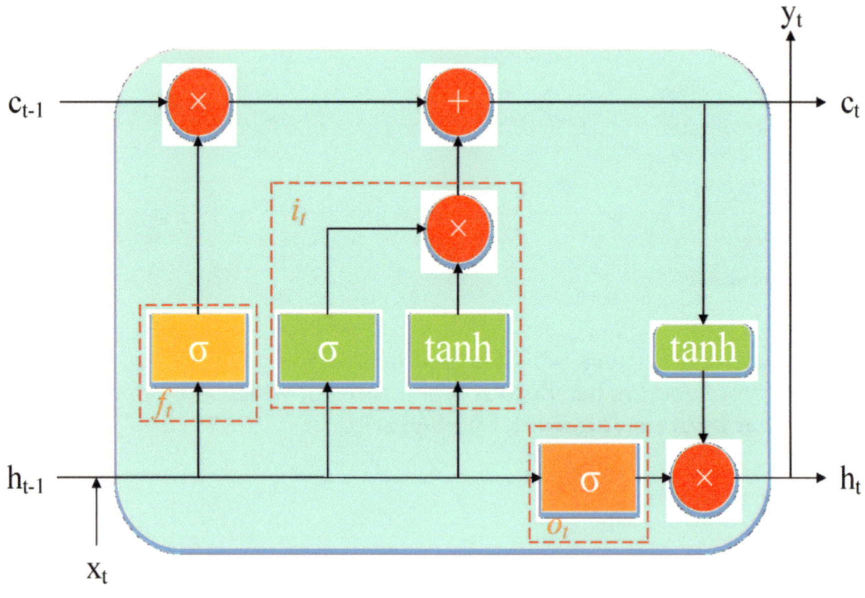

Figure 7.
LSTM base cell.

In Eqs. (6)-(13), i_t is the vector of the input gate, f_t is the vector of the output gate, and o_t is the vector of the forget gate. x_t is the input vector, y_t is the output vector, i_t, f_t, and o_t are the output of input gate, output gate, and forget gate, respectively; c_t and $\bar{c}_t$ are the finishing state and temporary of the memory cell in the memory block; σ is the sigmoid function which determines the propagation of the information; W_{xf}, W_{xi}, W_{xc}, and W_{xo} are the input weight matrices; W_{hf}, W_{hi}, W_{hc}, and W_{ho} are the reurrent weight matrices; W_{hy} is the hidden output weight matrix; b_f, b_i, b_c, b_o, and b_y are the related bias vectors.

8. Prediction based on ensemble learning

To overcome the vulnerability inherent in individual models, ensemble learning methods have been developed. These methods combine some basic models and produce a more robust predictor. The purpose of ensemble learning is to increase the accuracy and efficiency of the final model. While the results produced by individual models may be strong in one dataset and weak in another, the results produced by ensemble models consistently show greater accuracy than individual models. Ensemble learning methods not only improve prediction accuracy, but also reduce variance and increase robustness against anomalous data. Ensemble learning techniques can be divided into three main categories: Bagging, Boosting, and stacking. The stacking ensemble model is a single meta-learner model that usually includes two types of learners: basic and meta learners. The base learners can be heterogeneous and homogeneous methods of any type, and the learner can be a model such as linear regression, decision tree, neural networks, or others. The selection of meta learner strongly depends on the characteristics of basic learners.

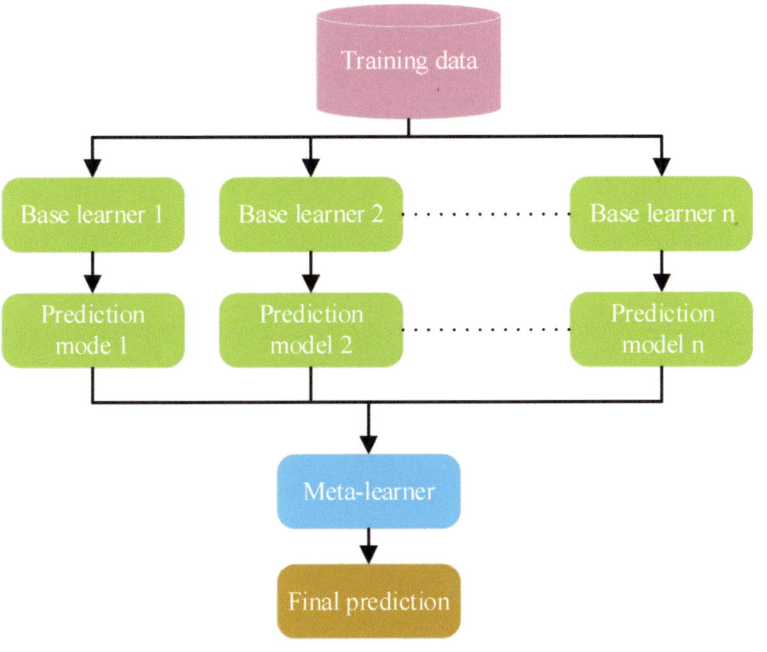

Figure 8.
The general structure of stacking ensemble model.

The stack method is a popular ensemble learning technique for improving estimation or classification. In the first step of batch learning, basic learners individually predict the training data. In the next step, the predictions made by the basic learners are used as training data for super-learning. Finally, the learner makes predictions and the overall accuracy of the model is estimated using the test data. The general structure of stack group learning is shown in **Figure 8**.

9. Conclusion

Unexpected behaviors of renewable distributed generations affect the modern power systems' planning, operating, and reliability. Also, the uncertain generation seriously affect the generators economy. Therefore, in this chapter, the models and techniques for accurate and robust prediction function have been proposed and discussed. The proposed high accuracy models decrease DG generation uncertainty prediction in energy system management and energy markets.

In this chapter, at the first, the effect of feature selection on prediction accuracy by using signal decomposition functions like wavelet transform has been discussed. Then, recent efficient methods for training and forecast based on neural networks have been explained. For more training capabilities, the deep learning-based methods such as long short-term memory and ensemble learning have been proposed. The results of the proposed methods present the abilities and accuracy for predicting the uncertain parameters in the modern power systems.

Author details

Farshid Keynia[1*] and Golamreza Memarzadeh[2]

1 Department of Energy Management and Optimization, Institute of Science and High Technology and Environmental Sciences, Graduate University of Advanced Technology, Kerman, Iran

2 Department of Electrical Engineering, Faculty of Engineering, Vali-e-Asr University of Rafsanjan, Rafsanjan, Iran

*Address all correspondence to: keynia@yahoo.com

References

[1] Afshar K, Shamsini Ghiasvand F, Bigdeli N. Optimal bidding strategy of wind power producers in pay-as-bid power markets. Renewable Energy. 2018;**127**:575-586

[2] Meng A, Ge J, Yi H, Chen S. Wind speed forecasting based on wavelet packet decomposition and artificial neural networks trained by crisscross optimization algorithm. Energy Conversion and Management. 2016;**114**:75-88

[3] Zhang W, Qu Z, Zhang K, Mao W, Ma Y, Fan X. A combined model based on CEEMDAN and modified flower pollination algorithm for wind speed forecasting. Energy Conversion and Management. 2017;**136**:439-451

[4] Wang J, Wang Y, Jiang P. The study and application of a novel hybrid forecasting model – A case study of wind speed forecasting in China. Applied Energy. 2015;**143**:472-488

[5] Ait Maatallah O, Achuthan A, Janoyan K, Marzocca P. Recursive wind speed forecasting based on Hammerstein auto-regressive model. Applied Energy. 2015;**145**:191-197

[6] Torres JL, García A, De Blas M, De Francisco A. Forecast of hourly average wind speed with ARMA models in Navarre (Spain). Solar Energy. 2005;**79**:65-77

[7] Yang D, Sharma V, Ye Z, Lim LI, Zhao L, Aryaputera AW. Forecasting of global horizontal irradiance by exponential smoothing, using decompositions. Energy. 2015;**81**:111-119

[8] Barbounis TG, Theocharis JB. A locally recurrent fuzzy neural network with application to the wind speed prediction using spatial correlation. Neurocomputing. 2007;**70**:1525-1542

[9] Focken U, Lange M, Mönnich K, Waldl HP, Beyer HG, Luig A. Short-term prediction of the aggregated power output of wind farms – A statistical analysis of the reduction of the prediction error by spatial smoothing effects. Journal of Wind Engineering and Industrial Aerodynamics. 2002;**90**:231-246

[10] Wang J, Du P, Niu T, Yang W. A novel hybrid system based on a new proposed algorithm-multi objective whale optimization algorithm for wind speed forecasting. Applied Energy. 2017;**208**:344-360

[11] Sun W, Liu M. Wind speed forecasting using FEEMD echo state networks with RELM in Hebei, China. Energy Conversion and Management. 2016;**114**:197-208

[12] Niu T, Wang J, Zhang K, Du P. Multi-step-ahead wind speed forecasting based on optimal feature selection and a modified bat algorithm with the cognition strategy. Renewable Energy. 2017;**118**:213-229

[13] Liu H, Tian H, Li Y. Four wind speed multi-step forecasting models using extreme learning machines and signal decomposing algorithms. Energy Conversion and Management. 2015;**100**:16-22

[14] Santamaría-Bonfil G, Reyes-Ballesteros A, Gershenson C. Wind speed forecasting for wind farms: A method based on support vector regression. Renewable Energy. 2016;**85**:790-809

[15] Sfetsos A. A comparison of various forecasting techniques applied to mean hourly wind speed time series. Renewable Energy. 2000;**21**:23-35

[16] Khodayar M, Kaynak O, Khodayar ME. Rough deep neural architecture for short term wind speed forecasting. IEEE Transactions on Industrial Informatics. 2017;**13**(6): 2770-2779

[17] Hu Q, Zhang R, Zhou Y. Transfer learning for short-term wind speed prediction with deep neural networks. Renewable Energy. 2016;**85**:83-95

[18] Wang HZ, Li GQ, Wang GB, Peng JC, Jiang H, Liu YT. Deep learning-based ensemble approach for probabilistic wind power forecasting. Applied Energy. 2017;**188**:56-70

[19] Kuremoto T, Kimura S, Kobayashi K, Obayashi M. Time series forecasting using a deep belief network with restricted Boltzmann machines. Neurocomputing. 2014;**137**:47-56

[20] Guo Z, Zhou K, Zhang X, Yang S. A deep learning model for short-term power load and probability density forecasting. Energy. 2018;**160**:1186-1200

[21] Chen K, Chen K, Wang Q, He Z, Hu J. Short-term load forecasting with deep residual networks. IEEE Transactions on Smart Grid. 2019;**10**:3943-3952

[22] Keynia F, Memarzadeh G. A new financial loss/gain wind power forecasting method based on deep machine learning algorithm by using energy storage system. IET Generation, Transmission & Distribution. 2022;**16**(5):851-868

[23] Memarzadeh G, Keynia F. Solar power generation forecasting by a new hybrid cascaded extreme learning method with maximum relevance interaction gain feature selection. Energy Conversion and Management. 2023;**298**(4):117763

[24] Amjady N, Keynia F. A new prediction strategy for price spike forecasting of day-ahead electricity markets. Applied Soft Computing. 2011;**11**(6):4246-4256

[25] Memarzadeh G, Keynia F. A new short-term wind speed forecasting method based on fine-tuned LSTM neural network and optimal input sets. Energy Conversion and Management. 2020;**213**:112824

[26] Amjady N, Keynia F. Day-ahead price forecasting of electricity markets by mutual information technique and cascaded neuro-evolutionary algorithm. IEEE Transactions on Power Apparatus and Systems. 2009;**24**(1):306-318

[27] Kouhi S, Keynia F. A new cascade NN based method to short-term load forecast in deregulated electricity market. Energy Conversion and Management. 2013;**71**:76-83

[28] Heydari A, Garcia D, Keynia F, Bisegna F, De Santoli L. A novel composite neural network-based method for wind and solar power forecasting in microgrids. Applied Energy. 2019;**251**:113353

Section 2

Renewable Energy Sources and Applications

Chapter 3

Insight into Hydrogen: The Future's Most Promising Energy Carrier in Preparation, Storage, Transportation, and Applications

Tiande Mo, Yong Lyu, Faheem Mushtaq and Hong Zhao

Abstract

The problem of climate change is highly critical at present, as extreme weather events triggered by greenhouse gases are occurring more frequently. Reducing the concentration of greenhouse gases in the atmosphere has become an urgent matter. Developing clean energy is a crucial means to achieve "carbon neutrality" and ultimately "climate neutrality". Hydrogen energy is one of the promising major energy sources of the future. This chapter will discuss the basic concepts of hydrogen energy, including its classification, production, transportation, storage, and applications. In particular, the focus is on green hydrogen, which is produced through electrolysis using renewable energy sources such as wind and solar power, as green hydrogen is considered the most sustainable form of hydrogen. Accordingly, the potential applications of hydrogen, such as fuel cells, hydrogen engines, the preparation of e-methanol, and green ammonia, are also discussed. The aim of this review paper is to help the public understand and accept hydrogen energy and realize the significance of reducing carbon emissions in the transition to a low-carbon economy.

Keywords: climate change, carbon neutrality, clean energy, hydrogen energy, fuel cell

1. Introduction

In recent years, the global energy landscape has been undergoing a profound transformation, motivated by the pressing necessity to combat climate change and decrease greenhouse gas emissions [1, 2]. The concept of carbon neutrality has become a central theme in energy policies worldwide, which involves balancing the carbon emitted with the carbon absorbed from the atmosphere by utilizing carbon sinks. This ambitious goal necessitates transitioning from fossil fuels to renewable and low-carbon energy sources. Hydrogen energy has emerged as a promising solution in the quest for carbon neutrality [3]. Hydrogen energy refers to the use of hydrogen as a carrier for storing and delivering energy. Unlike traditional fossil fuels, hydrogen itself is not an energy source but an energy carrier [4]. It can be derived from various feedstocks like water, natural gas, and biomass using different methods, such as

Types	Source/Method	CO_2 emissions
Gray hydrogen	Natural gas/methanol steam reforming	Higher carbon
Blue hydrogen	Natural gas/methanol steam reforming with CO_2 capture	Lower carbon
Green hydrogen	Water/electrolysis	Zero carbon

Table 1.
Different types of hydrogen depending on their production source.

electrolysis, steam methane reforming, and gasification, ensuring versatility in production. One of the most significant advantages of hydrogen energy is its potential for zero-emission energy production. When hydrogen is used in fuel cells or combusted with oxygen, the sole byproduct is water, positioning it as an eco-friendly alternative to traditional fuels. This clean energy attribute positions hydrogen as a critical player in reducing greenhouse gas emissions and combating climate change [5].

Hydrogen is categorized into different types based on its production sources, with each type color-coded to reflect its greenhouse gas emission profile, as shown in **Table 1** [6, 7].

1.1 Gray hydrogen

The most prevalent kind of hydrogen produced now is gray hydrogen. A procedure known as steam methane reforming (SMR) is a process that is made from natural gas. In this method, natural gas is heated with steam to create hydrogen and carbon dioxide. The subsequent atmospheric release of carbon dioxide makes gray hydrogen a substantial source of greenhouse gas emissions. Gray hydrogen is largely employed in industrial processes like petroleum refining and ammonia production [8].

1.2 Blue hydrogen

The same SMR method used to create gray hydrogen is also used to create blue hydrogen but with the addition of carbon capture and storage (CCS) technology. This innovation reduces greenhouse gas emissions by capturing the carbon dioxide created during the SMR procedure and storing it underground. Blue hydrogen is seen as a transitional fuel since it lowers emissions while enabling the continuing use of current infrastructure. Despite increased interest in using it for power generation and transportation, its primary use is in industrial applications [9].

1.3 Green hydrogen

"Green hydrogen" is produced when the energy required for electrolysis comes from renewable sources like wind, solar, and geothermal energy [10]. This process is generally emissions-free and offers a means to store renewable energy for future use. Currently, it is not cost-effective to liquefy green hydrogen. Therefore, it may be combined with nitrogen to form green ammonia or with CO_2 to produce green methanol. These substances, known as energy carriers, can be used as e-fuels for transportation and storage or can be reconverted into green hydrogen. Green hydrogen is produced *via* electrolysis, a method that involves using electricity to split water into hydrogen and oxygen. The electricity employed in this process can be sourced from renewable energies such as wind and solar power, making green hydrogen an entirely emissions-free fuel. Currently, the production of green hydrogen is relatively costly, but as renewable energy becomes more prevalent, its cost is anticipated to decline. Green hydrogen holds the potential for diverse applications, such as in transportation, power generation, and heating.

In addition, there are "brown hydrogen" and "white hydrogen". Brown hydrogen is created by burning coal to produce hydrogen. Because coal is a high-carbon fossil fuel, the production of brown hydrogen leads to substantial carbon emissions, which have negative effects on the environment [11]. Known as "white hydrogen", it can be generated through various geological processes [12]. The study of geologic hydrogen and its potential as an energy resource is an active area of research, as it holds promise for renewable energy applications, particularly in the context of hydrogen fuel cells and clean energy production. It is important to note that the creation of geologic hydrogen is generally a slow and long-term process, occurring over geological timescales. This is because the other methods are human production technology methods, and this is created by a natural phenomenon. The availability and abundance of geologic hydrogen can vary significantly depending on the specific geological setting and the interplay of various factors such as rock composition, temperature, pressure, and the presence of suitable reactants.

2. Production of hydrogen

2.1 Methane reforming for hydrogen production

Methanol steam reforming for hydrogen production was developed by Johnson-Matthey in the 1970s [13]. In the industrial methanol reforming process, methanol and desalinated water serve as raw materials, undergoing a catalytic reforming reaction at temperatures between 220 and 280°C. The process achieves a methanol conversion rate of over 99% in a single pass, with hydrogen selectivity exceeding 99.5%. The reformed gas contains only trace amounts of methane and carbon monoxide, besides hydrogen and carbon dioxide. By using pressure swing adsorption, hydrogen with a purity of 99.999% can be obtained, with carbon monoxide content below 5 ppm. The process of hydrogen production through methane reforming is a reversible endothermic reaction, as shown in **Figure 1**, generally conducted at high temperatures. During

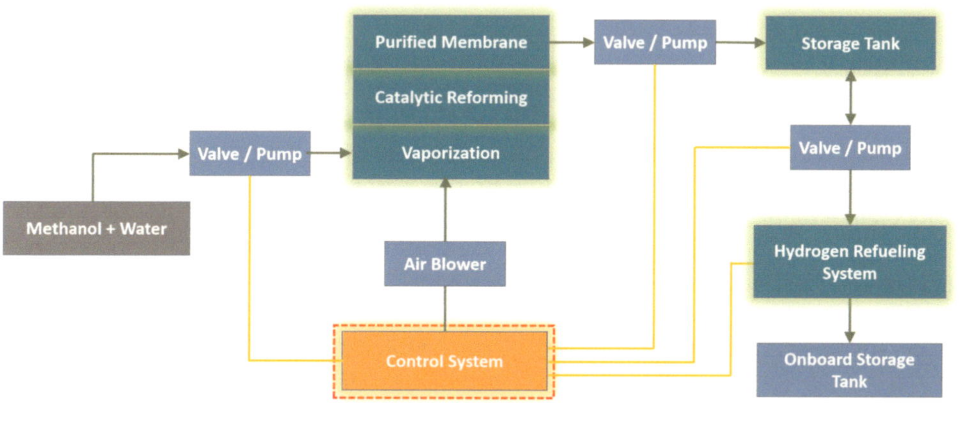

Figure 1.
Flowchart of methane reforming for hydrogen production.

this reaction, it is accompanied by the water-gas shift reaction of CO. The catalyst commonly used is Ni/Al_2O_3, often with additives to inhibit carbon deposition on the catalyst. To accelerate the removal of carbon from the catalyst surface, excess steam can be used in the reaction feed. The synthesis gas obtained from methane steam reforming enters the water-gas shift reactor, where, through high and low-temperature shift reactions, CO is converted into CO_2 and additional H_2, thereby increasing the hydrogen yield [14]. The reaction equations are shown below:

$$CH_4 + H_2O = CO + 3H_2\ (\Delta H = 205.8\,\text{kJ/mol}) \tag{1}$$

$$CO + H_2O = CO_2 + H_2\ (\Delta H = -41.2\,\text{kJ/mol}) \tag{2}$$

$$CH_4 + 2H_2O = CO_2 + 4H_2 \tag{3}$$

$$CH_4 + CO_2 = 2CO + 2H_2 \tag{4}$$

$$CH_4 + 3CO_2 = 4CO + 2H_2O \tag{5}$$

$$CH_4 = C + 2H_2 \tag{6}$$

$$2CO = C + CO_2 \tag{7}$$

$$CO + H_2 = C + H_2O \tag{8}$$

$$CO_2 + 2H_2 = C + 2H_2O \tag{9}$$

$$CH_4 + 2CO = 3C + 2H_2O \tag{10}$$

$$CH_4 + CO = 2C + 2H_2O \tag{11}$$

Although this set of reactions does not represent the reaction mechanism, it is consistent with known chemical processes. In practical processes, since water vapor is always in excess, the carbon deposition reactions (6)–(11) are avoided. Based on the experimental research and analysis by Hou et al. [15], the reaction rates of Eqs. (4) and (5) are very slow. From a chemical kinetics standpoint, these two reactions can be overlooked in thermodynamic studies of the process. Eq. (3) is the result of the superposition of Eqs. (1) and (2), so what actually determines the reaction equilibrium are Eqs. (1) and (2), namely the reforming reaction and the shift reaction. The shift reaction further converts CO to generate H_2 on the basis of the reforming reaction, making the reforming reaction the key reaction among the two. In the SMR process, the main process operating parameters include temperature, pressure, steam-to-carbon ratio, and space velocity. Process operations cannot be considered in isolation; their impact on the reaction must be estimated along with factors such as catalysts, economic costs, and materials. This method is highly efficient and cost-effective, making it the dominant method for hydrogen production globally. However, it has significant environmental drawbacks, primarily the emission of CO_2, a greenhouse gas contributing to global warming. To mitigate these effects, carbon capture and storage (CCS) technologies are being integrated into the SMR process [16].

2.2 Hydrogen production from petroleum byproducts

Coke oven gas is a byproduct of the coke production process, with approximately 420 Nm^3 of coke oven gas produced per ton of coke. The composition of coke oven gas includes 55–60% hydrogen by volume, 23–27% methane, and 6–8% carbon monoxide. After removing impurities such as naphthalene and sulfur, the hydrogen in coke oven gas can be purified using pressure swing adsorption devices [17]. In the caustic soda industry, a significant amount of hydrogen is produced as a byproduct during the electrolysis of saltwater to produce caustic soda. Although most large chlor-alkali plants are equipped with hydrochloric acid and polyvinyl chloride units to balance chlorine and recycle byproduct hydrogen, only about 60% is recovered for the production of hydrochloric acid, vinyl chloride monomer, and hydrogen peroxide [18]. The remaining hydrogen, aside from a small amount compressed at hydrogen compression stations and transported in cylinders, is mostly used as boiler fuel [19]. Propane dehydrogenation and light hydrocarbon cracking are also important options for fuel cell hydrogen sources.

2.3 Hydrogen production *via* water electrolysis

Electrolytic water hydrogen production is the cleanest and most sustainable method of producing hydrogen and is set to become one of the most promising hydrogen production methods in the development of fuel cells [20, 21]. It is an important technological means to achieve a sustainable hydrogen energy economy. By the end of 2023, the European Union Clean Energy Technology Observatory (CETO) has published the "Water Electrolysis and Hydrogen in the European Union - 2023 Status Report on Technology Development, Trends, Value Chains and Markets". This document provides a concise analysis of the current status of hydrogen production technologies in the EU, the emerging trends, demands for downstream applications, and the potential challenges confronting the related market. The most environmentally friendly method to produce hydrogen is through the use of renewable energy sources, like wind or solar power, to drive electrolysis. This process employs an electric current to separate water into hydrogen and oxygen. By using renewable energy sources, the carbon emissions associated with hydrogen production and combustion are nearly zero (the diagram is shown in **Figure 2**). The only possible emissions may come from indirect sources, such as hydrogen storage or transportation. Advancing green hydrogen production technologies is essential for transitioning to a low-carbon economy and satisfying the increasing demand for clean energy.

Water electrolysis is a clean and sustainable method for producing hydrogen. It involves the decomposition of water (H_2O) into hydrogen (H_2) and oxygen (O_2) using an electric current [22]. The process takes place within an electrolyzer, comprising an anode and a cathode divided by an electrolyte. Upon application of an electric current, water molecules at the anode decompose into oxygen, protons, and electrons. The protons traverse the electrolyte to the cathode, where they unite with electrons to generate hydrogen. The overall reaction for water electrolysis is shown in **Table 2.** This method is highly sustainable, especially when the electricity consumed is produced from renewable sources such as wind, solar, or hydropower. It produces pure

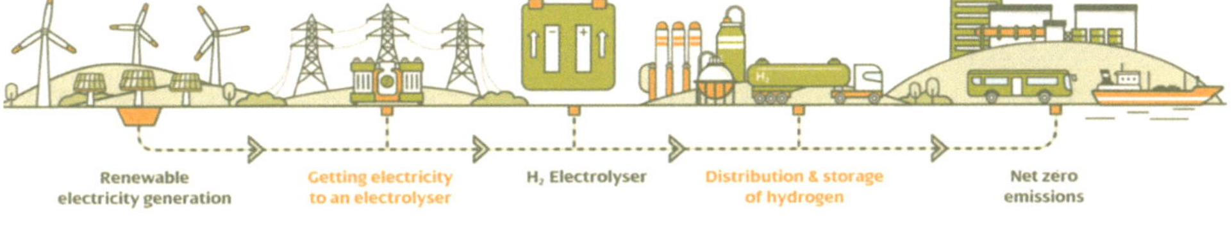

Figure 2.
Green hydrogen production and utilization (https://energytracker.asia/hydrogen-production-methods/).

Oxygen evolution in acid			
$H_2O \Leftrightarrow OH^{\bullet} + e^- + H^+$ **(1)**			Standard electrode potential φ^{θ}
$OH^{\bullet} \Leftrightarrow O^{\bullet} + e^- + H^+$	$O^{\bullet} + O^{\bullet} \Leftrightarrow O_2$		1.229 V
Oxygen evolution in alkaline			**Pt|O$_2$|H$^+$**
$OH^- \Leftrightarrow OH^{\bullet} + e^-$ **(2)**		acid electrolyte $\Delta E^{\theta} = 1.23$ V	
$OH^- + OH^{\bullet} \Leftrightarrow O^{\bullet} + H_2O + e^-$	$O^{\bullet} + O^{\bullet} \Leftrightarrow O_2$		0.401 V
Hydrogen evolution in acid			**OH$^-$|O$_2$|Pt**
$H_3O^+ + e^- \Leftrightarrow H^{\bullet} + H_2O$**(3)**			
$H_3O^+ + e^- + H^{\bullet} \Leftrightarrow H_2O$	$H^{\bullet} + H^{\bullet} \Leftrightarrow H_2$		0.000 V
Hydrogen evolution in alkaline			**H$^+$|H$_2$|Pt**
$H_2O + e^- + H^{\bullet} + OH^-$ **(4)**		alkaline electrolyte $\Delta E^{\theta} = 1.23$ V	
$H_2O + e^- + H^{\bullet} \Leftrightarrow H_2 + OH^-$	$H^{\bullet} + H^{\bullet} \Leftrightarrow H_2$		−0.8277 *V*
			OH$^-$|H$_2$|Pt

Table 2.
Standard potential for water splitting under different pH conditions [23].

hydrogen without any carbon emissions, which makes it an appealing choice for attaining carbon neutrality. However, the current cost of electrolysis is relatively high, primarily due to the expense of electricity and the materials used in electrolyzers. Efforts in research and development are continuously being made to boost the efficiency and decrease the cost of this technology.

Electrolysis of water primarily includes five major technologies as shown in **Figure 3** top: alkaline electrolysis (a), anion exchange membrane electrolysis (AEM) (b), proton exchange membrane electrolysis (PEM) (c), high-temperature protonic-ceramic electrolysis cell (PCEC). (d) High-temperature solid-oxide electrolysis (SOEC) (e) [24, 25]. Among these, alkaline electrolysis and proton exchange membrane electrolysis are relatively mature, while high-temperature solid-oxide electrolysis and anion exchange membrane electrolysis are developing slowly. Proton-conducting ceramic electrolysis is at an earlier stage of development. The currently most commercially mature technology is alkaline water electrolysis, using a 20–30 wt% concentrated NaOH or KOH solution as the electrolyte. However, alkaline solutions easily react with CO_2 in the air to produce K_2CO_3, resulting in high ohmic resistance losses, which typically lead to operation at relatively low current densities (200–400 $mAcm^{-2}$). Additionally, alkaline water electrolyzers struggle to respond to transient loads, which poses problems when paired with renewable power, and they also have high material corrosion rates [26].

Proton exchange membrane (PEM) electrolyzers have higher proton conductivity and can operate at lower temperatures and higher current densities [27]. However, the perfluoropolymer PEM membranes required for PEM electrolyzers are often costly, and the catalysts needed are expensive platinum and iridium noble metals, costing three to four times more than alkaline water hydrogen production. Anion exchange membrane water electrolysis combines the simplicity and ease of operation of solid polymer electrolytes with low cost and can use alkaline-compatible electrodes and hydrocarbon membranes, attracting widespread attention. Tsinghua University in

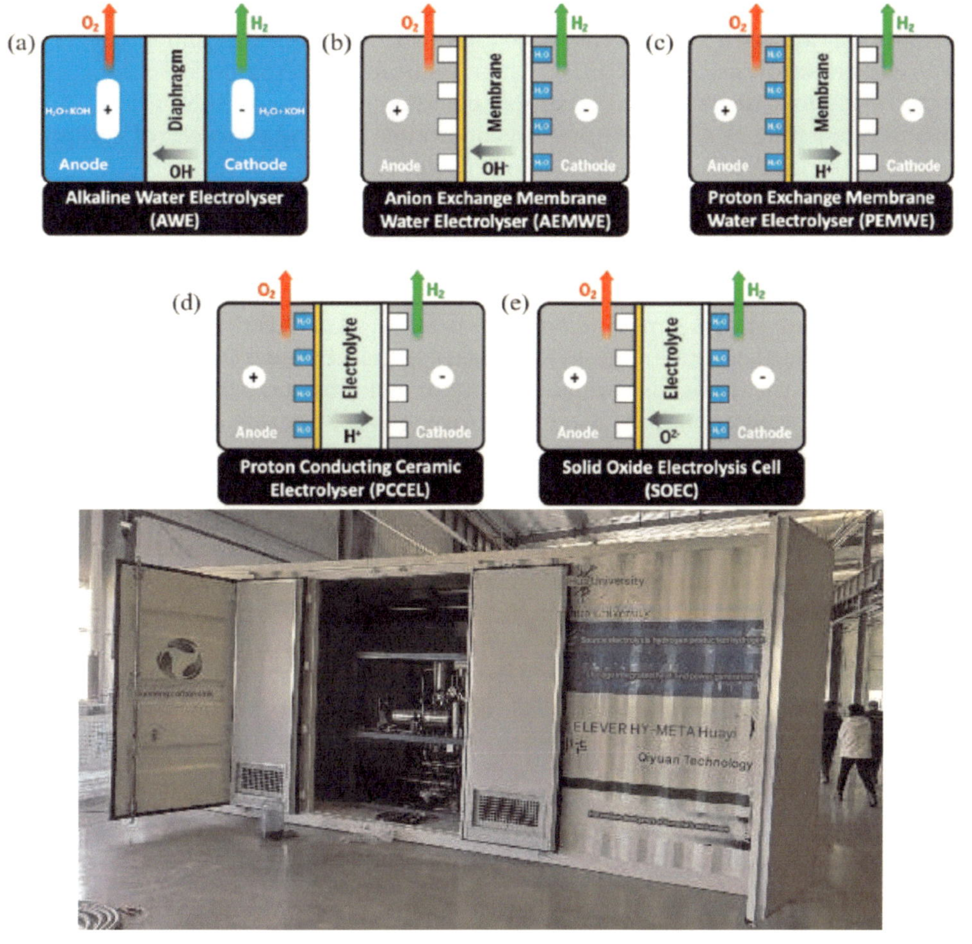

Figure 3.
Different types of electrolytic cells and their principles (top) [24], a demo of a PEM electrolyzer developed by Tsinghua University (bottom).

China has established a large-scale PEM hydrogen production demonstration facility, as shown in **Figure 3**, bottom.

Protonic ceramics represent an innovative class of materials with promising applications in various intermediate-temperature electrochemical processes, such as the generation of "green" hydrogen. Renewable electricity can be employed to drive steam electrolysis, resulting in the production of H_2 and O_2 within a protonic-ceramic electrolysis cell (PCEC) [28]. Operating typically between 400 and 600°C, water vapor carried by an air stream is introduced into a gas channel adjacent to the air–steam electrode, known as the "positrode". These gases diffuse through the porous positrode to the electrode-electrolyte interface, where H_2O is electrochemically split into protons and molecular oxygen. The protons travel through the protonic-ceramic electrolyte to the fuel electrode, or "negatrode", where it recombines to form H_2. Molecular hydrogen diffuses through the porous negatrode into the neighboring fuel chamber before exiting the cell. These devices have the potential to generate pure, dry, pressurized, carbon-free hydrogen using water and renewable electricity sources. Proton-conducting ceramics exhibit a low activation energy for proton transport [29].

This facilitates high proton conductivity at operating temperatures that are potentially 200–300°C lower than those of the more established oxygen-ion conducting (O^{2-}) solid-oxide electrolysis cells (SOECs). These reduced temperatures help mitigate thermally induced degradation processes, including phenomena like the agglomeration and coarsening of nickel, chromium contamination, and the creep of components in metallic stacks. Additionally, PCECs offer operational flexibility, potentially functioning effectively and efficiently across various steam concentrations [30].

Among the current international hydrogen production technologies, medium to high-temperature electrolysis using SOEC is more efficient, achieving an efficiency of over 85%, and has entered the demonstration and validation stage [31]. For SOECs, the solid-oxide electrolysis method employs ceramic materials that conduct oxygen ions (O^{2-}) as the electrolyte [32]. Water enters the electrolyzer in the form of steam, where it is electrolyzed at the cathode into hydrogen ions and oxygen ions. The hydrogen ions receive electrons conducted through an external circuit to form hydrogen gas, while the oxygen ions migrate through the ceramic electrolyte to the anode, generating oxygen gas and releasing electrons. This electrolysis method must operate in a high-temperature environment (650–800°C). Under the high-temperature and high-humidity conditions of electrolysis, the commonly used Ni/YSZ hydrogen electrode is prone to oxidation, resulting in a loss of activity. The performance degradation mechanism and microstructure regulation still require further research. Conventional oxygen electrodes have issues with anode polarization and delamination in electrolysis mode, necessitating the development of new materials and new oxygen electrodes to reduce polarization losses. Additionally, an important development project involves evaluating and testing the electrolysis characteristics and heat absorption and dissipation properties of solid-oxide electrolyzer cell SOEC. The advantages of SOEC technology include high energy conversion efficiency and the absence of the need for precious metal catalysts. It can also directly produce syngas from steam and CO_2 for various applications, such as the synthesis of liquid fuels [32].

2.4 Extraction of underground hydrogen

Underground hydrogen extraction is an emerging method that involves tapping into natural hydrogen reserves found in geological formations [33]. These reserves can be found in various forms, such as hydrogen-rich gas fields, underground caverns, and in association with other minerals and hydrocarbons. The extraction process typically involves drilling wells and using techniques similar to those employed in natural gas extraction. There are several advantages to this method. Firstly, it can potentially provide a large and continuous supply of hydrogen without the need for complex chemical processes. Secondly, it may offer a lower carbon footprint compared to conventional hydrogen production methods, depending on the geological context and the extraction techniques used. However, this method is still in the early stages of development and faces several challenges. These include the identification and characterization of viable hydrogen reservoirs, the development of efficient extraction technologies, and the assessment of environmental impacts. Further research and exploration are necessary to unlock the potential of underground hydrogen extraction.

3. Hydrogen storage

Hydrogen storage is a critical aspect of the hydrogen energy ecosystem. Effective storage solutions are essential for the practical and widespread use of hydrogen as an energy carrier. This chapter delves into the various methods of hydrogen storage, each with its unique advantages and challenges.

3.1 High-pressure hydrogen storage

High-pressure hydrogen storage involves compressing hydrogen gas to high pressures, typically between 350 and 700 bar. This method is widely used due to its relatively straightforward technology and ease of implementation. High-pressure storage systems are commonly found in hydrogen refueling stations and fuel cell vehicles. Compressing hydrogen gas increases its energy density, making it more efficient for storage and transportation. Also, high-pressure tanks and cylinders have been extensively researched and developed, ensuring safety and reliability. Currently, compressing hydrogen to high pressures requires significant energy, which can reduce the overall efficiency of the hydrogen energy cycle. Hence, high-pressure systems must be meticulously designed and maintained to prevent leaks and ensure safe operation.

3.2 Liquid hydrogen storage

Liquid hydrogen storage involves cooling hydrogen gas to cryogenic temperatures (around −253°C or − 423°F) to convert it into a liquid state [34]. This method significantly increases the energy density of hydrogen, making it suitable for applications where space is limited. Liquid hydrogen has a much higher energy density compared to its gaseous form, facilitating more compact storage solutions. Liquid hydrogen can be transported in insulated tanks, making it feasible for long-distance distribution. However, maintaining hydrogen in a liquid state requires continuous cooling, which consumes energy and adds to operational costs. Further, over time, some hydrogen will evaporate (boil off) due to heat ingress, leading to potential losses and safety concerns.

3.3 Solid-state hydrogen storage

Solid-state hydrogen storage involves absorbing hydrogen into solid materials, such as metal hydrides, which can store hydrogen at a molecular level [35]. This method offers a safe and compact way to store hydrogen without the need for high pressures or cryogenic temperatures. Solid-state storage eliminates the risks associated with high-pressure gas and cryogenic liquids. Metal hydrides and other materials can store hydrogen densely, making them ideal for applications with space constraints [36]. But the materials used for solid-state storage can be expensive and may require rare or specialized elements. It should be mentioned that hydrogen release from solid-state materials can be slower compared to other storage methods, potentially limiting the rate at which hydrogen can be utilized.

So far, material-based hydrogen storage is an emerging field that explores innovative materials, such as carbon nanotubes, metal-organic frameworks (MOFs) [37], and other advanced composites, to store hydrogen more efficiently [38]. Advanced materials possess the potential to store significant quantities of hydrogen at lower pressures and temperatures. Material properties can be tailored to optimize hydrogen absorption and release rates. However, many material-based storage solutions are still in the experimental stage and require further research to become commercially viable.

And developing and manufacturing these advanced materials can be costly, potentially limiting their widespread adoption.

3.4 Hydrogen underground storage and the hydrogen system

Underground storage will play a pivotal role in any expansive hydrogen economy, with the future hydrogen network potentially operating similarly to the existing natural gas network. A pipeline network will connect hydrogen supply to a wide range of customers, each with varying scales and demand requirements. Underground hydrogen storage will be an essential tool for operators, allowing them to regulate hydrogen flow to satisfy customer demands at any time. It will enable operators to balance supply and demand, maintain infrastructure integrity by ensuring pressure and temperature remain within design limits, guarantee uninterrupted supply as per contractual obligations, protect against supply or demand disruptions, and postpone some additional investments in electricity infrastructure. Underground hydrogen storage serves two main purposes: (1) meeting energy demand during peak periods and (2) compensating for low renewable energy generation. Similar to the natural gas system, maintaining a balance between supply and demand will be necessary across various timeframes—hourly, daily, weekly, and seasonally. The significance of underground gas storage will become even more critical than its current role, ensuring the overall resilience of the energy system. While natural gas systems can partially depend on other storage solutions like liquefied and compressed aboveground storage, such methods are not viable for large-scale hydrogen storage applications.

4. Applications of hydrogen energy

Hydrogen energy is increasingly being recognized for its potential to revolutionize various industries. This chapter delves into the current applications of hydrogen energy, highlighting its versatility and the transformative impact it is having across different sectors.

The "Big Hydrogen Energy" concept: Firstly, leveraging the energy attribute of hydrogen to replace traditional energy sources through fuel cells or direct combustion; secondly, utilizing hydrogen's role as a carrier to connect hydrogen energy with renewable energy, thereby enhancing the consumption capacity of renewable energy; and thirdly, employing hydrogen's raw material attribute to replace traditional fossil energy-based hydrogen production processes, achieving significant carbon reduction. The examples of end-use scenarios for green hydrogen are shown in **Figure 4**. We will outline the primary applications of hydrogen in the sections that follow.

4.1 Fuel cells

When hydrogen is generated in the form of molecular hydrogen, its energy can be released by combining with oxygen to form water [40]. This can be achieved using traditional internal combustion engines or devices known as fuel cells. In a fuel cell, hydrogen energy is efficiently converted directly into electricity with minimal power loss. As such, hydrogen acts as an energy carrier, facilitating the transport, storage, and delivery of energy derived from other sources. Fuel cells represent one of the most significant applications of hydrogen energy. These devices convert hydrogen gas into electricity through a chemical reaction with oxygen, with water being the only

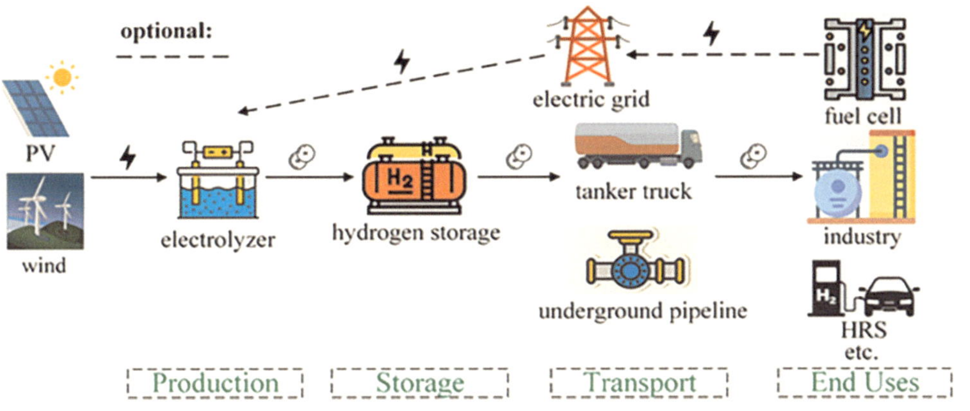

Figure 4.
Examples of end-use scenarios for green hydrogen [39].

Hydrogen Storage Capability, Mileage, and Availability				
Hydrogen storage (35Mpa)	~0.1 kg	~1.5 kg	~5 kg	~10 kg
Mileage	~120km	~10 hours	~800km	~600km

Figure 5.
Fuel cell vehicles and their hydrogen range.

byproduct. This makes fuel cells an environmentally friendly alternative to conventional fossil fuels. They are being used in a variety of applications, including motorcycles, forklifts, cars, and busses, as shown in **Figure 5**, top. Their hydrogen storage and driving range are also summarized in **Figure 5**, bottom. Hydrogen fuel cell vehicles (FCVs) are increasingly popular due to their longer driving ranges and quicker refueling times compared to battery-electric vehicles. Major automakers like Toyota, Honda, and Hyundai have already introduced FCVs to the market. Fuel cells are also used for stationary power generation, providing reliable and clean energy for residential, commercial, and industrial applications. They are particularly useful in remote locations where grid connectivity is limited. Portable Power: Small fuel cells are being developed for portable electronic devices, offering a longer lifespan and quicker recharge than traditional batteries.

4.2 H_2 internal combustion engine (ICE)

In addition to fuel cell-electric and battery-electric applications, the hydrogen engine stands as a pivotal powertrain technology for the future [41]. Bosch has developed hydrogen direct injection to fulfill the demands of applications for both on-highway and off-highway, covering a broad spectrum of performance and engine uses in heavy commercial vehicles. With the mixture being produced inside a closed cylinder, the direct injection of hydrogen effectively removes the risk of backfiring into the intake tract. Additionally, this technology allows for multiple injections, which improve the combustion process. Another advantage is that by optimizing the engine for hydrogen direct injection with appropriate operational strategies, it is feasible to attain performance and driving characteristics similar to traditional engines. Direct injection technology can adapt to various hydrogen supply qualities and engine operating conditions, thereby increasing the adaptability and flexibility of the fuel. This means H_2 ICE can more effectively respond to diverse driving conditions and environmental requirements, delivering stable and reliable power output for vehicles. Additionally, direct injection aids in forming a stratified charge within the combustion chamber. By implementing direct injection, multiple injections can occur within one engine cycle to create a stratified charge, supporting the advancement of ultra-dilute stratified combustion technology in this project.

However, in a hydrogen ICE, the high-temperature combustion results in the production of harmful nitrogen oxides (NOx), similar to other internal combustion engines [42]. Therefore, water vapor is not the sole emission from the tailpipe of a hydrogen-fueled internal combustion engine. With the precise and repeatable injection capabilities provided by direct injection, an air-fuel mixture concentration gradient can be established inside the combustion chamber. The most combustible air-fuel mixture, with an equivalence ratio near stoichiometric, is controlled to form near the center between the injector and piston. It is proven that hydrogen combustion under stratified charge can fully exploit the fuel's broad flammability range, while the low combustion temperature can suppress NOx formation. Implementing a two-stage burning process can increase burning speed, thereby enhancing thermal efficiency [43].

4.3 Steelmaking

The direct reduction of iron involves the chemical extraction (reduction) of oxygen from iron ore in its solid state [44]. Presently, the steelmaking process reduces iron from iron ore chemically using fossil resources like natural gas or coal. During this process, carbon reacts with the oxygen in the iron ore, yielding metallic iron and a carbon-rich process gas, as illustrated by the simplified chemical reaction below:

$$2Fe_2O_3 + 3C = 4Fe + 3CO_2$$

Iron ore can be reduced using hydrogen as a replacement for carbon, resulting in water as the waste gas, according to the following reactions:

$$Fe_2O_3 + 3H_2 = 2Fe + 3H_2O, FeO + H_2 = Fe + H_2O$$

Hydrogen is currently produced from fuels rich in hydrogen, such as natural gas and biogas, as well as through water electrolysis. Presently, natural gas is the predominant source, contributing to approximately 75% of the global hydrogen production, which totals around 70 million tonnes annually. This represents about 6% of the world's natural gas consumption. At present, water electrolysis contributes to less than 0.1% of the world's hydrogen production. In the process of producing direct reduced iron with natural gas, hydrogen is utilized in the reduction process, though it is combined with carbon. The greenhouse gas emissions from gas-based production of direct reduced iron are lower than those from the Blast Furnace method, with each ton of direct reduced iron generating 1.5 tons of CO_2 emissions. Currently, pure hydrogen is not utilized in ironmaking processes. Some innovative process routes, such as carbon capture and storage at blast furnaces, smelting reduction, and gas-based direct reduction method, are anticipated to

DOI: http://dx.doi.org/10.5772/intechopen.1010550

cost 10–50% more than existing commercial alternatives, depending on the regional context. This cost increase far exceeds current profit margins in the steel industry.

However, risk assessments and controls must be updated to account for risks related to hydrogen usage. Like other energy carriers, hydrogen carries certain health and safety risks when utilized on a large scale. Hydrogen's small molecular size allows it to permeate some materials, such as specific iron and steel pipes, potentially increasing their risk of failure. It also tends to escape more easily through seals and connectors than larger molecules like natural gas. Moreover, hydrogen has the potential to cause embrittlement and cracking in steel pipes and vessels. Steel manufacturers are currently developing and implementing processes to mitigate risks associated with the containment loss of hazardous, toxic, or flammable materials.

4.4 Methanol production

Global methanol production reaches about 90 million tons each year, with the majority (approximately 65%) derived from natural gas using steam methane reforming, while the remaining 35% is obtained from coal through gasification processes [45, 46]. The urgent demand for low-carbon fuels during the energy transition has significantly increased interest in producing methanol from renewable energy sources, a concept introduced by Nobel laureate—George A. Olah in 2005. Presently, the production of renewable (green) methanol worldwide is minimal, less than 0.2 million tons annually (about 0.2% of total production), mainly derived from biomass gasification [47]. Recently, a variety of advanced processes and technologies have been proposed for generating renewable methanol from sources like municipal solid waste, sludge, and others. The rising adoption of intermittent renewable energy sources like solar and wind, along with the demand for sustainable fuels to reduce carbon emissions in "hard-to-abate" sectors, especially in industries and heavy transportation, is making the production of methanol from renewable electricity (e-methanol) and captured carbon dioxide more attractive. CO_2 hydrogenation, generally performed in a fixed-bed catalytic reactor at temperatures between 250 and 300° C and pressures of 50–100 bar, is presently the technology showing the most promising potential for commercial development [48]. In essence, Renewable electricity is harnessed to produce green hydrogen *via* water electrolysis, with CO_2 being captured from concentrated sources such as flue gas from power plants or industrial facilities, or directly from the atmosphere. The power-to-methanol strategy provides a dual advantage: it facilitates the storage of renewable electricity in a chemical form and produces a fully renewable product that can replace methanol derived from fossil fuels. E-methanol and its derivatives offer appealing alternatives for the transportation sector, as they are exceptionally clean and compatible with the current distribution and storage infrastructure for conventional oil-based transport fuels.

4.5 Ammonia production

Ammonia is generally synthesized *via* the catalytic reaction between nitrogen and hydrogen. Although technological advancements have been made over the years, the core chemical process remains identical to the method established by Haber and Bosch in the early twentieth century [49]:

$$N_2\ (g) + 3\ H_2\ (g) = 2\ NH_3\ (g)\ \Delta H = -92\ \text{kJ/mole}\ (-46\ \text{kJ/mole for 1 mole of}\ NH_3)$$

The reaction typically occurs over iron catalysts at temperatures of approximately 400–600°C and pressures ranging from 200 to 400 atmospheres. By using hydrogen derived from renewable sources, green ammonia can be produced, significantly reducing the carbon footprint. This green ammonia can also serve as a carbon-free fuel or hydrogen carrier, thereby broadening its applications. Nevertheless, the potential use of liquid anhydrous ammonia, NH_3, as a hydrogen storage medium onboard vehicles or as a distribution medium has received little attention. When compared to the reforming of hydrocarbon fuels, extracting or cracking hydrogen from ammonia can be a less complicated process. The process results in nitrogen as a byproduct, which can be safely released from the vehicle without harming the environment, thus negating the need for a recycling system required in some other chemical hydride systems. Nonetheless, ammonia possesses certain chemical and physical properties that must be carefully considered when designing and engineering an onboard storage system. These considerations include its high thermal expansion coefficient, high vapor pressure under ambient conditions, its tendency to react with water, its reactivity with container materials, and the high toxicity of its vapor. These factors will be discussed in more detail below. It is crucial to note that PEM fuel cells cannot tolerate ammonia, even in very low concentrations, meaning that an onboard ammonia-based storage system would require a cracking reactor to prevent the passage of undissociated ammonia and/or a highly efficient filtration system [50].

Beyond the major applications mentioned above, it is being explored as a potential fuel for rockets and aircraft, offering a high energy density and clean combustion. Also, hydrogen can be burned directly to provide heat and power for industrial processes and residential heating, offering a low-carbon alternative to natural gas. Therefore, the applications of hydrogen energy are diverse and expanding. With technological advancements and the realization of economies of scale, hydrogen is set to become a key player in the worldwide shift toward a sustainable energy future.

5. Application predictions and successful implementation of H_2 considering environmental impact/low-carbon options

As shown in **Figure 6**, with the right conditions, we found that hydrogen could become a cost-effective low-carbon solution in 22 applications by 2030. These include long-distance transport and regional trains, which are notably competitive compared to other low-carbon options, as shown by their high positioning on the y-axis. Although these hydrogen applications collectively represent up to 15% of global energy consumption (17,500 TWh), it does not mean hydrogen will fulfill all this energy demand by 2030. Rather, it highlights hydrogen's potential as a key player in the future clean energy landscape. When measured against traditional alternatives, numerous applications stand out as highly competitive at scale, demanding either zero or low-carbon pricing for hydrogen to be viable, as reflected by their placement on the right side of the x-axis. This is particularly true for nine applications, such as trucks, trains, long-range passenger vehicles, and long-distance busses. On the flip side, other uses like turbines, industrial feedstock, or synthetic aviation fuels would need a carbon tax of at least USD 100 per ton of CO_2 equivalent to make hydrogen a cost-effective choice against conventional fuels.

Our analysis indicates that, despite the potential for elevated costs, hydrogen can effectively compete with the leading low-carbon alternative, battery-electric vehicles (BEVs), particularly for long-distance travel segments. Unlike other sectors, mobility

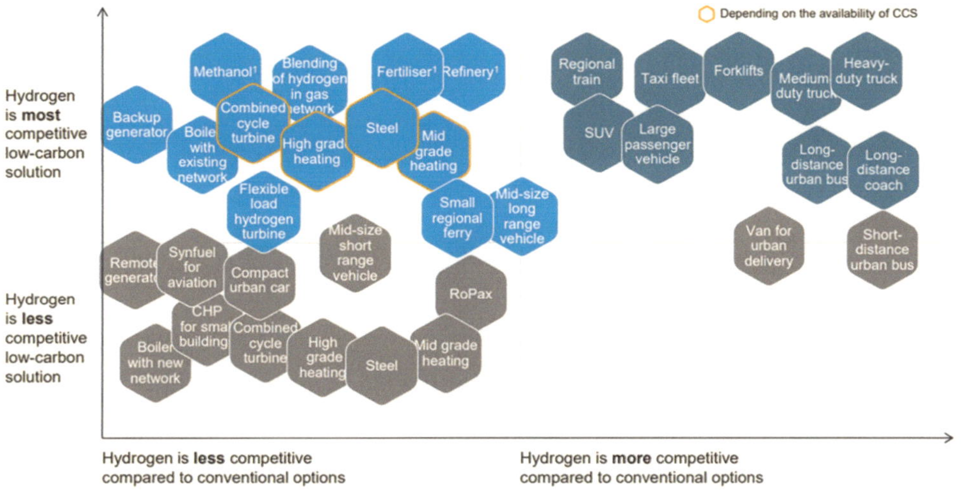

Figure 6.
Application of H_2 based on environmental factors/low-carbon solution. (Refer to: Path to hydrogen competitiveness: A cost perspective 20 January 2020).

applications are generally less impacted by hydrogen production costs due to the extensive nature of hydrogen supply chains and the significant cost of equipment. As a result, the hydrogen industry is well-positioned to fulfill a growing portion of transportation energy demand, even if production costs exceed USD 2 per kilogram, before factoring in supply chain and refueling expenses.

By 2030, hydrogen is expected to satisfy a substantial portion of the energy needs in the mobility sector. As shown in **Figure 7**, at a price of USD 6 per kilogram at the pump, covering production, distribution, and retail, hydrogen could cost-effectively fulfill around 15 percent of transportation energy requirements by 2030. We anticipate that this cost structure will be practical across various regions and applications by then. If the cost were reduced to USD 4 per kilogram at the point of use, hydrogen could potentially meet over 50 percent of the energy demand within the mobility sector. Trucks, long-haul busses, and larger passenger vehicles are particularly well-suited to benefit, given the prohibitive battery costs required to achieve an equivalent range with battery-powered alternatives.

In 2020, Foshan successfully deployed 1000 hydrogen fuel cell busses. These busses have collectively traveled over 90 million kilometers, making Foshan one of the largest hubs for hydrogen-powered public transportation in China. The city boasts 36 hydrogen refueling stations, the most of any city in the nation. The promotion of hydrogen fuel cell busses has significantly contributed to the growth of Foshan's hydrogen energy sector. Presently, around 170 enterprises and projects have been established in the area with an estimated total investment exceeding 65 billion yuan, yielding impressive results. To date, Foshan has introduced over 1600 hydrogen-powered vehicles across various categories, including busses, passenger cars, logistics vans, wing vans, dump trucks, and sanitation vehicles. Busses account for approximately 60% of these vehicles. The widespread adoption of hydrogen busses has heightened Foshan's reputation in the hydrogen energy industry, positioning it as a leading city. Foshan has collaborated with cities in five provinces, including Guangzhou, Shenzhen, and Fuzhou, to successfully apply for the first national demonstration city cluster for fuel cell vehicles (**Figure 8**, upper).

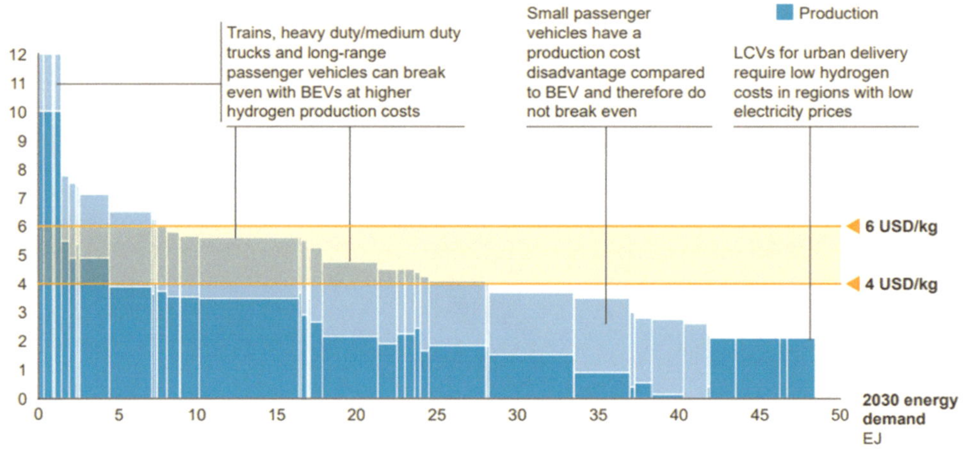

Figure 7.
Application predictions of H_2 based on cost (refer to: Path to hydrogen competitiveness A cost perspective 20 January 2020).

In Hong Kong, the first hydrogen fuel cell double-decker bus began service in early 2024, and the initial three hydrogen-powered street cleaning vehicles are set to embark on an 18-month trial later in the year. Additionally, the MTR Corporation plans to conduct non-passenger tests of hydrogen-powered trams by 2024 to assess the feasibility of implementing this technology in Hong Kong. Discussions are

Figure 8.
Successful implementation of H_2 in Foshan city, China, and Hong Kong, China (refer to https://td.gd.gov.cn/ztzl/jtqgjszl/cgzs/content/post_4458133.html and https://www.info.gov.hk/gia/general/202406/17/P2024061700427.htm).

underway with the transport industry to introduce other hydrogen-fueled heavy vehicles suitable for local use (**Figure 8**, bottom).

6. Conclusions and perspectives

In summary, advancing and integrating hydrogen energy, particularly green hydrogen, are pivotal actions toward achieving carbon neutrality and confronting global climate challenges. Green hydrogen, produced from renewable energy sources such as wind and solar power, represents a sustainable and powerful solution for reducing carbon emissions. Its potential applications across various sectors, such as transportation, industrial processes, and power generation, highlight its versatility and significance in the shift to a low-carbon economy.

The importance of green hydrogen transcends mere emission reductions; it signifies a fundamental transformation in our energy perception and usage. Unlike fossil fuels, which heavily contribute to greenhouse gas emissions, green hydrogen provides a clean alternative that can be produced and consumed without environmental damage. This positions it as a crucial element in strategies aimed at mitigating climate change impacts and fulfilling the objectives set by international accords such as the Paris Agreement.

Looking forward, the broad adoption of green hydrogen necessitates collective efforts in research, development, and infrastructure investment. Promoting public understanding and acceptance of hydrogen energy is essential for its successful deployment. Education and awareness initiatives can play a crucial role in dispelling myths and misconceptions about hydrogen safety and its environmental advantages. Furthermore, governments can ease this transition by instituting policies that encourage the adoption of hydrogen technologies through tax credits, subsidies, and research and development grants.

Policymakers, industry leaders, and researchers need to collaborate to surmount existing challenges such as cost, storage, and transportation to unlock the full potential of hydrogen energy. Technological innovations, including enhanced electrolyzers, improved storage solutions, and efficient distribution networks, are vital to making hydrogen a feasible and competitive energy source. Moreover, international cooperation can expedite progress by sharing knowledge, resources, and best practices across borders.

As we advance, the global commitment to green hydrogen can propel innovation, generate new economic opportunities, and contribute to a cleaner, more sustainable future. By prioritizing the advancement of hydrogen technologies, we lay the groundwork for a resilient energy system capable of meeting the world's increasing energy needs while protecting the environment. Embracing green hydrogen not only addresses present energy challenges but also establishes a foundation for future generations to inhabit a world where clean, sustainable energy is the standard rather than the exception.

Acknowledgements

The author thanks the support of the project of Guangdong provincial basic and applied research fund: 2023A1515140176.

The author acknowledges the usage of ChatGPT and DeepSeek for language polishing of the manuscript.

Author details

Tiande Mo[1], Yong Lyu[1], Faheem Mushtaq[2] and Hong Zhao[1*]

1 Hong Kong Productivity Council, Hong Kong

2 CIC EnergiGUNE Basque, Spain

*Address all correspondence to: mezhaohong@gmail.com

References

[1] Hassan Q, Viktor P, Al-Musawi TJ, Ali BM, Algburi S, Alzoubi HM, et al. The renewable energy role in the global energy transformations. Renewable Energy Focus. 2024;**48**:100545

[2] Gielen D, Gorini R, Wagner N, Leme R, Gutierrez L, Prakash G, et al. Global energy transformation: A roadmap to 2050. 2019:1-49

[3] Evro S, Oni BA, Tomomewo OS. Carbon neutrality and hydrogen energy systems. International Journal of Hydrogen Energy. 2024;**78**:1449-1467

[4] Züttel A, Remhof A, Borgschulte A, Friedrichs O. Hydrogen: The future energy carrier. Philosophical Transactions of the Royal Society A: Mathematical, Physical and Engineering Sciences. 2010;**368**(1923):3329-3342

[5] Hassan Q, Algburi S, Sameen AZ, Salman HM, Jaszczur M. Green hydrogen: A pathway to a sustainable energy future. International Journal of Hydrogen Energy. 2024;**50**:310-333

[6] Howarth RW, Jacobson MZ. How green is blue hydrogen? Energy Science & Engineering. 2021;**9**(10):1676-1687

[7] Saha P, Akash FA, Shovon SM, Monir MU, Ahmed MT, Khan MFH, et al. Grey, blue, and green hydrogen: A comprehensive review of production methods and prospects for zero-emission energy. International Journal of Green Energy. 2024;**21**(6):1383-1397

[8] Zainal BS, Ker PJ, Mohamed H, Ong HC, Fattah I, Rahman SA, et al. Recent advancement and assessment of green hydrogen production technologies. Renewable and Sustainable Energy Reviews. 2024;**189**:113941

[9] Oni A, Anaya K, Giwa T, Di Lullo G, Kumar A. Comparative assessment of blue hydrogen from steam methane reforming, autothermal reforming, and natural gas decomposition technologies for natural gas-producing regions. Energy Conversion and Management. 2022;**254**:115245

[10] Nnabuife SG, Hamzat AK, Whidborne J, Kuang B, Jenkins KW. Integration of renewable energy sources in tandem with electrolysis: A technology review for green hydrogen production. International Journal of Hydrogen Energy. 2024;**107**:218-240

[11] Lal A, You F. Targeting climate-neutral hydrogen production: Integrating brown and blue pathways with green hydrogen infrastructure via a novel superstructure and simulation-based life cycle optimization. AICHE Journal. 2023;**69**(1):e17956

[12] Joseph A, Oghenegare E. Recent advances in white hydrogen exploration and production: A mini review. Journal of Energy Research and Reviews. 2023;**13**(4):64-79

[13] Sheldon D. Methanol production-a technical history. Johnson Matthey Technology Review. 2017;**61**(3):172-182

[14] Angeli SD, Monteleone G, Giaconia A, Lemonidou AA. State-of-the-art catalysts for CH_4 steam reforming at low temperature. International Journal of Hydrogen Energy. 2014;**39**(5):1979-1997

[15] Hou D, Liang W, Zhang B, Li C. Seepage law of mixed gases in CO_2 displacement of coal seam CH_4. Journal of China Coal Society. 2019;**44**(11):3463-3471

[16] Soltani SM, Lahiri A, Bahzad H, Clough P, Gorbounov M, Yan Y. Sorption-enhanced steam methane reforming for combined CO_2 capture and hydrogen production: A state-of-the-art review. Carbon Capture Science & Technology. 2021;**1**:100003

[17] Ortiz-Imedio R, Ortiz A, Urroz J, Diéguez P, Gorri D, Gandía L, et al. Comparative performance of coke oven gas, hydrogen and methane in a spark ignition engine. International Journal of Hydrogen Energy. 2021;**46**(33): 17572-17586

[18] Abreu CM, Fonseca AC, Rocha NM, Guthrie JT, Serra AC, Coelho JF. Poly (vinyl chloride): Current status and future perspectives via reversible deactivation radical polymerization methods. Progress in Polymer Science. 2018;**87**:34-69

[19] Habib MA, Abdulrahman GA, Alquaity AB, Qasem NA. Hydrogen combustion, production, and applications: A review. Alexandria Engineering Journal. 2024;**100**:182-207

[20] Zhao H, Chen C, Chen D, Saccoccio M, Wang J, Gao Y, et al. $Ba_{0.95}La_{0.05}FeO_{3-\delta}$–multi-layer graphene as a low-cost and synergistic catalyst for oxygen evolution reaction. Carbon. 2015; **90**:122-129

[21] Wang J, Zhao H, Gao Y, Chen D, Chen C, Saccoccio M, et al. $Ba_{0.5}Sr_{0.5}Co_{0.8}Fe_{0.2}O_{3-\delta}$ on N-doped mesoporous carbon derived from organic waste as a bi-functional oxygen catalyst. International Journal of Hydrogen Energy. 2016;**41**(25): 10744-10754

[22] Zahra R, Pervaiz E, Yang M, Rabi O, Saleem Z, Ali M, et al. A review on nickel cobalt sulphide and their hybrids: Earth abundant, pH stable electro-catalyst for hydrogen evolution reaction. International Journal of Hydrogen Energy. 2020;**45**(46):24518-24543

[23] Lei Q, Wang B, Wang P, Liu S. Hydrogen generation with acid/alkaline amphoteric water electrolysis. Journal of Energy Chemistry. 2019;**38**:162-169

[24] Chatenet M, Pollet BG, Dekel DR, Dionigi F, Deseure J, Millet P, et al. Water electrolysis: From textbook knowledge to the latest scientific strategies and industrial developments. Chemical Society Reviews. 2022;**51**(11): 4583-4762

[25] Emam AS, Hamdan MO, Abu-Nabah BA, Elnajjar E. A review on recent trends, challenges, and innovations in alkaline water electrolysis. International Journal of Hydrogen Energy. 2024;**64**: 599-625

[26] Dash S, Singh A, Jose S, Elangovan D, Surapraraju SK, Natarajan SK. Advances in green hydrogen production through alkaline water electrolysis: A comprehensive review. International Journal of Hydrogen Energy. 2024;**83**:614-629

[27] Toghyani S, Afshari E, Baniasadi E, Atyabi S, Naterer G. Thermal and electrochemical performance assessment of a high temperature PEM electrolyzer. Energy. 2018;**152**:237-246

[28] Thatte AA. High temperature water splitting using protonic ceramic electrolysis cells: multiscale design analysis from cells to systems. Colorado School of Mines; 2020

[29] Herradon C, Le L, Meisel C, Huang J, Chmura C, Kim Y, et al. Proton-conducting ceramics for water electrolysis and hydrogen production at elevated pressure. Frontiers in Energy Research. 2022;**10**:1020960

[30] Su H, Hu YH. Degradation issues and stabilization strategies of protonic ceramic electrolysis cells for steam electrolysis. Energy Science & Engineering. 2022;**10**(5):1706-1725

[31] Ni M, Leung MK, Leung DY. Technological development of hydrogen production by solid oxide electrolyzer cell (SOEC). International Journal of Hydrogen Energy. 2008;**33**(9):2337-2354

[32] Zheng Y, Wang J, Yu B, Zhang W, Chen J, Qiao J, et al. A review of high temperature co-electrolysis of H_2O and CO_2 to produce sustainable fuels using solid oxide electrolysis cells (SOECs): Advanced materials and technology. Chemical Society Reviews. 2017;**46**(5): 1427-1463

[33] Epelle EI, Obande W, Udourioh GA, Afolabi IC, Desongu KS, Orivri U, et al. Perspectives and prospects of underground hydrogen storage and natural hydrogen. Sustainable Energy & Fuels. 2022;**6**(14):3324-3343

[34] Yanxing Z, Maoqiong G, Yuan Z, Xueqiang D, Jun S. Thermodynamics analysis of hydrogen storage based on compressed gaseous hydrogen, liquid hydrogen and cryo-compressed hydrogen. International Journal of Hydrogen Energy. 2019;**44**(31): 16833-16840

[35] Lee S-Y, Lee J-H, Kim Y-H, Kim J-W, Lee K-J, Park S-J. Recent progress using solid-state materials for hydrogen storage: A short review. PRO. 2022; **10**(2):304

[36] Klopčič N, Grimmer I, Winkler F, Sartory M, Trattner A. A review on metal hydride materials for hydrogen storage. Journal of Energy Storage. 2023; **72**:108456

[37] Zhao H, Sheng L, Wang L, Xu H, He X. The opportunity of metal organic frameworks and covalent organic frameworks in lithium (ion) batteries and fuel cells. Energy Storage Materials. 2020;**33**:360-381

[38] Qureshi F, Yusuf M, Ahmed S, Haq M, Alraih AM, Hidouri T, et al. Advancements in sorption-based materials for hydrogen storage and utilization: A comprehensive review. Energy. 2024;**309**:132855

[39] Zhang J, Li J. Revolution in renewables: Integration of green hydrogen for a sustainable future. Energies. 2024;**17**:4148

[40] Lubitz W, Tumas W. Hydrogen: an overview. Chemical Reviews. 2007; **107**(10):3900-3903

[41] Verhelst S, Wallner T. Hydrogen-fueled internal combustion engines. Progress in Energy and Combustion Science. 2009;**35**(6):490-527

[42] Nguyen D, Turner JW. Effect of nitric oxide on the performance and emissions of a hydrogen-fueled HCCI engine. Fuel. 2023;**349**:128611

[43] Szwaja S, Jamrozik A, Tutak W. A two-stage combustion system for burning lean gasoline mixtures in a stationary spark ignited engine. Applied Energy. 2013;**105**:271-281

[44] Ogbezode JE, Ajide OO, Oluwole OO, Ofi O. Recent trends in the technologies of the direct reduction and smelting process of iron ore/iron oxide in the extraction of iron and steelmaking. In: Iron Ores and Iron Oxides-New Perspectives. London, UK: IntechOpen; 2023

[45] Olah GA, Goeppert A, Prakash GS. Beyond oil and gas: The methanol economy. Angewandte Chemie

International Edition. 2005;**44**:2636-2639

[46] Sollai S, Porcu A, Tola V, Ferrara F, Pettinau A. Renewable methanol production from green hydrogen and captured CO_2: A techno-economic assessment. Journal of CO_2 Utilization. 2023;**68**:102345

[47] Goeppert A, Olah GA, Prakash GS. Toward a sustainable carbon cycle: The methanol economy. In Book: Green Chemistry. Elsevier; 2018. pp. 919-962

[48] Latsiou AI, Charisiou ND, Frontistis Z, Bansode A, Goula MA. CO_2 hydrogenation for the production of higher alcohols: Trends in catalyst developments, challenges and opportunities. Catalysis Today. 2023; **420**:114179

[49] Liu H. Ammonia synthesis catalyst 100 years: Practice, enlightenment and challenge. Chinese Journal of Catalysis. 2014;**35**(10):1619-1640

[50] Thomas G, Parks G. Potential Roles of Ammonia in a Hydrogen Economy: A Study of Issues Related to the Use Ammonia for on-Board Vehicular Hydrogen Storage. Vol. 23. US Government of US Department of Energy; 2006

Section 3

System Integration and Economic Analysis

Chapter 4

The Economics of Integrating Distributed Energy Resources into the Electricity System

Matthew E. Oliver

Abstract

This chapter provides a broad overview of current economic issues related to integrating distributed energy resources (DERs)—primarily solar photovoltaics (PV) and battery electric storage (BES)—into the electricity system, and the implications this has for electricity markets. Topics covered include (but are not limited to) generation intermittency, merit-order effects, the 'duck curve' problem, resource adequacy and the 'missing-money' problem, efficient investment in flexible generation and storage, implications for the electricity transmission and retail distribution systems, incentivizing adoption of DERs, and the solar rebound effect. The chapter's general aim is to provide a non-economics audience with a broad sense of the current state of economic research on these topics.

Keywords: rooftop solar photovoltaics, small-scale battery storage, intermittent generation, electricity markets, economics, policy

1. Introduction

The transition to a carbon-free, clean energy future continues to accelerate worldwide. Distributed energy resources (DERs) such as rooftop solar photovoltaics (PV) and battery electric storage (BES) have emerged as a clear leader among sustainable, decentralized electricity technologies. This is no accident; a variety of social, political, technological, regulatory, and market forces have converged to propel these technologies to the forefront of energy decarbonization efforts in developed and developing countries alike. Further, solar PV is now among the lowest cost electricity sources and is forecast to be a prominent source of global primary energy supply by 2050 [1]. Pairing rooftop solar PV with small-scale BES allows households to shift the times they consume PV-generated electricity, reduce consumption of grid-supplied electricity, or even disconnect from the grid altogether [2].

The defining characteristics of rooftop solar PV are that (i) it incurs zero marginal generation cost, and (ii) generation is intermittent. As penetration of rooftop solar PV increases, these characteristics will complicate system operations, particularly when viewed against the backdrop of electricity market restructuring that has occurred across much of the US and Europe over the past three decades. Zero marginal generation cost implies a shift in the electricity supply-demand equilibrium that benefits consumers but may reduce the profitability of traditional utilities and independent power producers. In the absence of storage, intermittent generation introduces a host of additional (and often hidden) costs with respect to electricity system operations and market efficiency, making the continued deployment of BES paired with PV imperative for the efficient management of PV generation going forward. As the market for DERs continues to expand in its share of the overall generation mix, these key characteristics pose significant challenges for electricity grid operators seeking simultaneously to maintain the crucial supply-demand balance, minimize system costs, and avoid system failures, while maintaining an equitable distribution of the resulting costs and benefits.

This chapter provides an overview of the basic economics of integrating distributed energy into the broader electricity system. I review the economic challenges posed by the dual problems of zero-marginal-cost generation and intermittency, primarily in the context of restructured electricity markets. Of crucial importance is that the economic effects cannot be decoupled from the engineering constraints—these challenges are inexorably linked. Throughout, I highlight some important implications for energy policies designed to promote adoption of DER technologies.

2. Cost comparison: DERs versus traditional generation

A primary reason DERs are poised to expand rapidly in the coming decades is that they are now highly competitive with traditional electricity generation technologies on a levelized cost of energy (LCOE) basis. **Table 1** presents the US Energy Information Administration's LCOE estimates per megawatt hour (MWh) (in constant 2021 US dollars) for new generation coming into service in 2027 and 2040. Solar PV without BES is the lowest cost generation technology, at an estimated LCOE of $33.83 per MWh in 2027, falling to $31.07 by 2040. Solar PV will thus be roughly 15% cheaper on a LCOE basis than the least expensive fossil fuel technology, combined-cycle natural gas (CCNG), by 2027, and nearly 30% cheaper by 2040. Compared to the other prominent renewable technology, wind, PV is now cheaper than onshore wind, where LCOE will remain stable at around $40/MWh, and far outperforms

Generation type	2027	2040
Coal	$82.61	$79.46
CCNG	$39.94	$44.05
Nuclear	$81.71	$80.20
Wind, onshore	$40.23	$39.63
Wind, offshore	$105.38	$98.01
PV	$33.83	$31.07
PV + BES*	$49.03	$45.54

*Assumes PV coupled with a four-hour battery storage system.

Table 1.
Estimated LCOE for different electricity generation technologies (2021 USD per MWh). Source: U.S. Energy Information Administration, Annual Energy Outlook 2022. All estimates are based on relevant capacity factors and include levelized capital costs, fixed O&M, variable costs, transmission costs, and applicable tax credits, but do not include carbon or other emissions taxes.

Innovation scenario*	2030	2040
Conservative	$3350.93	$3159.56
Moderate	$2976.20	$2232.15
Advanced	$2115.22	$1319.68

Assumes a cost recovery period of 30 years, includes cost of installation.

Table 2.
Estimated capital cost ($/kW) for a 5 kW/12.5kWh residential BES system under alternative technological innovation scenarios (2022 USD). Source: U.S. National Renewable Energy Laboratory (see: https://atb.nrel.gov/electricity/2024/residential_battery_storage).

offshore wind. When paired with BES, the LCOE of solar PV rises considerably, but is forecast to remain much lower than coal, nuclear, and offshore wind. By 2040, PV + BES will be roughly equivalent to CCNG on a LCOE basis. Note also that these estimates do not include carbon or other emissions taxes on fossil fuel generation, which are either in place or being considered in many countries around the world and make PV even more competitive relative to CCNG.

However, research and development may lead to significant cost declines for BES systems that are not reflected in the **Table 1** forecasts of LCOE for PV + BES. If the installed cost of BES systems were to fall significantly over the same period, this would further improve the competitiveness of PV + BES relative to traditional generation technologies. **Table 2** presents the US National Renewable Energy Laboratory's most recent capital cost estimates for a 5 kW/12.5kWh residential BES system in 2030 and 2040 under alternative technological innovation scenarios. The cost of such a system in 2020 was around $4000/kW. Under conservative assumptions about technological innovation in small-scale BES, capital costs will have declined by roughly 16.25% over the current decade but would decline over the next decade by a modest 5.7%. By contrast, in the moderate and advanced innovation scenarios, costs will have decreased by roughly one-quarter to nearly one-half by 2030, and by 45% to as much as 67% by 2040. This illustrates the potential of technological innovation to vastly improve the affordability of BES, expanding its market potential and viability as a DER technology.

3. Short-term grid management with DERs: Reliance on markets

Modern wholesale electricity markets rely on the principle of *merit-order dispatch* (MOD) to supply a given quantity of electricity at minimum cost, at each point in time. MOD refers to the dispatching of power to meet demand (load), starting with the lowest marginal cost power plant, bringing additional plants online in order of successively higher marginal costs [3]. **Figure 1** displays a simplified example of MOD with three power plants, whose constant marginal generation costs, c_i $(i = 1,2,3)$, are such that $c_1 < c_2 < c_3$. The maximum generating capacity of plant i is Q_i^{max}, and the total generation capacity available to the dispatcher is $Q^{total} = Q_1^{max} + Q_2^{max} + Q_3^{max}$. The arrangement of plant-level capacities in order of increasing marginal costs forms a simple electricity supply curve. Denote total power demand at time t as $\overline{Q}_t$, where $t = o$ is the *off-peak* period and $t = p$ is the *peak* period, such that $\overline{Q}_o < \overline{Q}_p$. Electricity

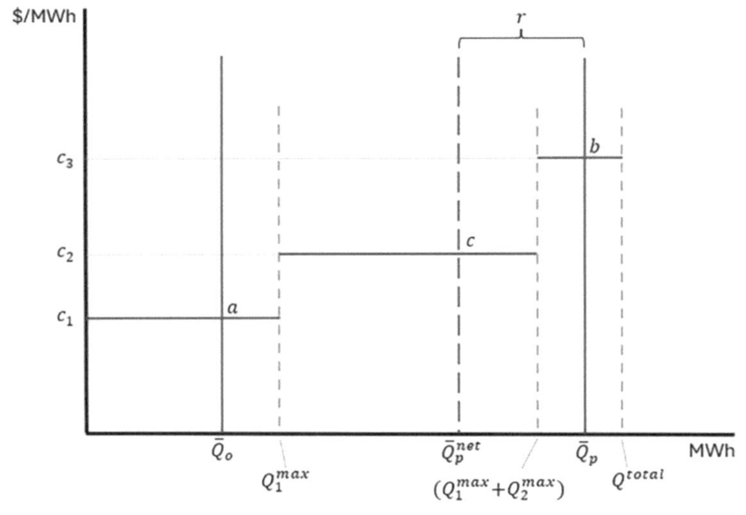

Figure 1.
Merit-order dispatch with three power plants. Off-peak and peak equilibria at points a and b, respectively. Addition of DERs during the peak period moves the peak-period equilibrium to point c.

demand at a given point in time is perfectly *price inelastic*, meaning it is independent of the price of electricity—hence, the demand line is vertical at $\overline{Q}_t$. During the off-peak period, demand intersects supply at point a, plant 1 supplies all power needed to meet demand $\overline{Q}_o$, and the equilibrium wholesale price is equal to the marginal cost of plant 1, $p_o^* = c_1$. Conversely, during the peak period, and in the absence of DERs, demand intersects supply at point b, plants 1 and 2 supply power up to their maximum capacities, and plant 3 supplies the remainder. The equilibrium wholesale price is equal to the marginal cost of plant 3, $p_p^* = c_3$.

Now consider how the introduction of DERs changes the market outcome. For simplicity, assume DERs only generate power during the peak period and denote the amount generated by DERs as r. Net demand on the power grid is now $\overline{Q}_p^{net} = \overline{Q}_p - r$, which intersects supply at point c. The peak-period equilibrium wholesale price falls to $p_p^{**} = c_2$. This reduction in the wholesale equilibrium price in the presence of DERs (as well as other zero- or low-marginal-cost generating capacity such as wind turbines) is known as the *merit-order effect* and has been observed empirically in markets with high penetration of solar (and wind) [4, 5].

Additional complications arise because generation from prominent DER technologies like solar PV is *intermittent*. While baseline fluctuations in solar PV generation due to the diurnal and seasonal cycles can be predicted with a fair degree of accuracy, they are still costly for system operators to account for. By contrast, random variation in meteorological factors (e.g., temperatures, precipitation, cloud cover) cause r to vary stochastically around its seasonal daily averages; such variation cannot be perfectly forecast. These combined sources of generation intermittency lead to additional variation in net demand, compounding the traditional demand variability from consumptive activities. Supply intermittency from DERs has several effects on electricity markets and system operations, such as increased variability of wholesale prices [6], episodes of negative prices [7, 8], increased need for costly storage and stand-by capacity [9, 10], larger and more frequent system imbalances [11], and greater reliance on costly ancillary services such as balancing, fast ramping, voltage control, and frequency control [12]. Generation intermittency accounts for a significant fraction of the total social costs of large-scale penetration of solar PV, although a much larger proportion is attributable to diurnal and seasonal intermittency than to unforecastable intermittency [13].

The challenges of generation intermittency from the diurnal cycle are perhaps best illustrated by the now famous *duck curve* phenomenon, named for the distinctive shape of the daily net load profile with high penetration of solar PV, shown for California in **Figure 2**. Duck curves have also been observed in other markets with high uptake of solar PV, including Germany and Australia. As generation from solar PV increases during the day, this reduces net load on the power grid. However, as the sun sets PV generation declines rapidly, and the timing of this drop-off in PV generation coincides with the daily evening peak load, causing net load to rise sharply. As aggregate solar generation capacity expands, the steepness of this evening ramp-up becomes more pronounced because the mid-day reduction in net load gets deeper.

Economists (and electricity systems researchers, generally) recognize that the most direct way to address the duck curve problem is through electricity storage. The complementarity between electricity storage and intermittent generation has featured prominently in recent economic research on electricity markets. Studies of investment, production, and storage in electricity systems with clean but intermittent technologies (e.g., solar) and polluting but dispatchable technologies (e.g., coal) have linked social efficiency to the design of economic support policies for clean energy investment [14, 15]. Others have examined combined solar and BES systems, where the optimal system size changes under different policy incentives and electricity pricing schemes [16]. A variety of modeling approaches and assumptions about market structure have been applied in studies of intertemporal electricity price arbitrage, to elucidate the conditions under which wholesale markets with a large proportion of intermittent generation capacity can provide adequate incentives for investment in BES [17–19]. The social efficiency of electricity price arbitrage through storage also depends on the degree of penetration of renewables and the emission rates of the marginal generators during peak and off-peak hours. In a market with low renewable penetration, if the emission rate of the off-peak marginal generator (e.g., coal) exceeds that of the peak-period marginal generator (e.g., natural gas), electricity price arbitrage increases pollution, reducing the net social benefit [20].

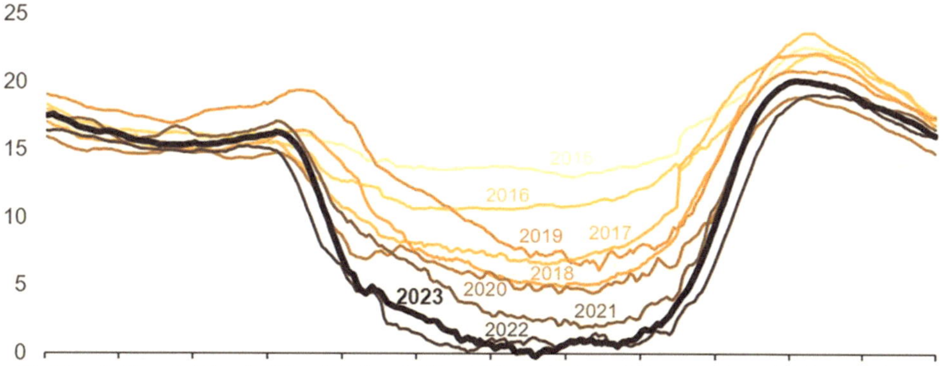

Figure 2.
Duck curve for California: CAISO lowest net load day each spring (March-May, 2015–2023), gigawatts. Source: US Enegy Information Administration (https://www.eia.gov/todayinenergy/detail.php?id=56880).

4. Long-term market challenges

Widespread diffusion of DER technologies like solar and BES presents significant long-term challenges for electricity markets and system operators. In this section, I first discuss electricity system reliability, the concepts of *resource adequacy* (RA) and the *missing-money problem* (MMP), and how these are affected by large-scale penetration of DERs. I then discuss how to incentivize efficient investment in flexible generation and BES to mitigate the problems caused by DER generation intermittency. Third, I explore some key issues related to transmission and local distribution. Fourth, I focus on retail electricity markets, particularly the possibility that rapid expansion of DERs could induce a so-called *utility death spiral*.

4.1 Reliability, resource adequacy, and the missing-money problem

RA refers to the ability of the electric system to meet aggregate electricity demand at all times. This concept is inherently related to *reliability*—a system will be reliable if its total installed capacity achieves RA (when operated within safety limits) [21]. In the short-run, one might expect that greater generation from DERs would improve RA because it reduces net electricity demand on the system (see **Figure 1**). However, RA is a long-run concept that is crucially related to investment in installed generation capacity, especially 'peaker' plants that sit idle for long periods of time and are fired only during brief spells of extreme peak demand, due to their high marginal operating costs. In restructured markets, such plants rely on these periods—when wholesale prices can spike to many multiples of the average—for the entirety of their annual revenue requirement. Yet, as shown above, the merit-order effect implies peak prices may be significantly reduced by widespread diffusion of DERs, leading to potentially large reductions in the revenues earned by peaker plants. This 'missing money' severely limits the profitability of investments in peaker plants. More generally, the MMP implies 'energy only' markets are unlikely to provide adequate revenues during peak demand periods to attract investment in new peaker plants (or prevent exit of existing ones), negatively impacting RA over the long run [22].

In vertically integrated markets, long-run capacity investment decisions are made as part of integrated electricity resource planning (IERP) programs. According to [23], IERP "is the process of identifying longer-term investments to meet electricity reliability requirements and public policy goals at a reasonable cost," and provides "a forum for regulators, electric utilities, and electricity industry stakeholders to evaluate the economic, environmental, and social benefits and costs of different investment options." In restructured markets, the independent system operator (ISO) is involved in IERP regarding investment in transmission capacity and ancillary services, but investments in generating capacity are made by independent power producers (IPPs) based on market price signals. If peak prices are insufficient to justify investment (due to merit-order effects), IERP practices—both in vertically integrated and restructured electricity markets—must be adapted to account for the increasing impact of DERs to better support RA and reliability. Instead of using IERP to determine how much DER generation to anticipate or procure, utilities and ISOs will increasingly be required to use separate forecasts of DER growth to adjust their residual resource requirements [23].

In restructured markets, another response to RA issues and the MMP has been the development of capacity markets. In its most basic form, a capacity market allows local distribution utilities to satisfy capacity requirements—typically expected peak monthly load plus a reserve margin—by paying IPPs for a 'recall right' that allows the ISO to recall them during extreme peak demand periods, thereby avoiding load curtailment [24]. In theory, capacity markets provide an efficient signal to investors to build capacity commensurate with consumer preferences for reliability and the *value of lost load* (VOLL), defined as the amount customers would pay to avoid a blackout [25]. Especially in a system with a cap on the wholesale electricity price set equal to VOLL, increasing penetration of intermittent DERs increases the need for a capacity payment because intermittent generation is an imperfect substitute for conventional, dispatchable generation, weakening its ability to mitigate system-cost externalities relative to its ability to satisfy electricity demand; in other words, greater intermittent DER penetration raises the social value of investment in peaker plants [26].

4.2 Investment in flexible generation and utility-scale storage

The duck curve problem and the increased variability of net load have led to greater need for flexible, or *fast-ramping*, generation technologies elsewhere on the system that can respond immediately to sudden changes in generation from intermittent DERs. Fast ramping is both costly and technologically challenging due to reduced fuel efficiency, accelerated depreciation due to physical strain on generator components, and the increased costs of specialized design characteristics needed for fast ramping to be possible [27]. Greater ramping costs represent another (often hidden) cost of intermittent DER generation. Given the need for greater investment in fast-ramping generation, some ISOs have introduced *flexible ramping products* (FRPs) into their real-time market operations to improve flexibility in responding to increased variability in net load [28]. An FRP is an ancillary service that, when priced correctly, provides an efficient long-run incentive to invest in fast-ramping generation [12].

A clear substitute for fast-ramping generation capacity is utility-scale electricity storage, which can typically be discharged quickly at low cost. Efficient long-run investment in storage capacity, however, comes with its own challenges. Cost projections show that installed cost per kWh of utility-scale lithium-ion BES is declining rapidly but is still significantly more expensive per unit capacity than its automotive and small electronics counterparts [29]. Despite this, utility-scale BES may be a more efficient approach than distributed BES, given that capital costs and market power are key determinants of the efficient scale of storage investment [30, 31]. Ultimately, electricity system managers may find it optimal to follow an all-of-the-above approach to investment in electricity storage technologies, pursuing both newer technologies like batteries, as well as investing in more mature technologies like fuel cells and pumped hydro, each of which has its own specific technical characteristics with respect to charge/discharge duration, energy density, and round-trip efficiency. Ultimately, if BES is to substitute for generation capacity over the long-run, the efficient level of BES investment must balance the returns from intertemporal price arbitrage against the precautionary need to ensure availability of supply during periods of peak net demand [18].

4.3 Impacts on transmission and distribution systems

There is widespread agreement among energy economists and policymakers that if deep decarbonization of the electricity system is to be achieved using current technologies, and within the time frame needed to avoid the most damaging consequences of global climate change, significant expansion and upgrade of the world's electricity transmission and distribution systems will be required [32, 33]. The most productive sites for utility-scale wind and solar are often located far from demand centers, requiring not only increases in transmission capacity, but also an expansion of the geographic area of interconnected wholesale market and dispatch regions to take advantage of dispersion effects to mitigate the system-wide costs of localized intermittency. One advantage of DERs like rooftop solar PV and on-site BES is that they are co-located with consumption and local demand, reducing the need for long-distance transmission. On the other hand, greater penetration of DERs will require significant investments by local utility providers to ensure these resources are integrated into the grid efficiently. For example, requiring smart-metering devices to be installed with residential PV and/or BES systems greatly enhances the utility's ability to efficiently integrate households' generation, consumption, and storage decisions into both short-term operation and long-term planning decisions [34, 35].

4.4 Implications for retail electricity markets

The rise of DERs has prompted some analysts to predict an impending *utility death spiral*, prompted by concerns that rapid expansion of DERs threatens the traditional business model of retail electricity providers [36, 37]. The logic is as follows. As DER adoption increases, net electricity demand falls, forcing utilities to raise rates to recover fixed costs from continually shrinking energy sales. Higher retail rates accelerate DER adoption, further reducing net electricity demand, inducing additional rate increases, and so on until the utility is driven out of the market entirely [38]. The conditions under which a utility death spiral can (or will) occur in a given market setting depend crucially on a complex interaction of factors including utility cost structure, PV and BES adoption rates, and net metering compensation [39].

Other researchers have been more conservative, arguing that utility death spirals, while theoretically possible, are unlikely to occur. At current rates of PV adoption and retail rate increases, concerns about utility death spirals may be premature at best and grossly overstated at worst [40]. Even with widespread adoption of DER technologies, widespread disconnection from the utility grid may not be a realistic projection for retail markets. Although increased penetration of DER systems will continue to reduce net demand, retail markets are not entirely static in their approach to ratemaking. Retail rate structures can be amended to help utilities develop other sources of revenue rather than increase volumetric prices [41]. Utilities can respond to financial distress in a variety of ways, including price-discrimination in favor of more price-elastic customers, redesign of retail rates and other cost recovery mechanisms to mitigate the risk of financial insolvency, or seeking direct regulatory protection for reasons of public or political interest [42].

5. Consumer behavior

In this section, I discuss two important topics for ongoing research regarding the economics of DERs, related to consumer behavior. First, I discuss the economics of the decision to adopt DERs and the variety of economic and financial incentive schemes policymakers have developed to encourage adoption. Second, I discuss the evidence that adoption of residential solar PV systems leads to a *rebound effect*, in which households increase total electricity consumption after installing solar PV.

5.1 Incentivizing adoption of DERs

The decision to adopt distributed energy technologies like solar PV and/or BES is complex and multi-faceted. Empirical studies have identified a range of factors that play a role, including socioeconomic and demographic factors, home size, access to financing, and solar compensation policies [43–45]. Adoption depends on a combination of two-part electricity tariffs (i.e., marginal prices and fixed fees), compensation schemes for excess generation, and installation cost subsidies, where each has its own implications for economic efficiency and distributional equity [46, 47]. Relying on installation cost subsidies generally results in a lower aggregate welfare loss but may be regressive, whereas using compensation schemes and marginal prices to incentivize adoption is more cost efficient for utilities and more progressive across the income distribution, but generates a larger welfare loss. Other studies have used real-options approaches to model the adoption of residential PV and BES. Optimal timing of investment is contingent on favorable market conditions, and the ability to store energy in a battery increases the household's optimal PV system size [48]. A more flexible approach treats the adoption of PV and battery systems as separate but linked compound options [49].

The net cost of installation is a major determinant of rooftop solar PV and battery adoption, prompting a variety of subsidy schemes to reduce net installation costs. Perhaps counterintuitively, however, generous installation subsidies have been shown to increase equilibrium PV system prices [50]. It is well known that up-front rebates, income tax credits, and other direct cash incentives, which serve to reduce net installation costs, have a large, positive effect on residential solar installations [51–55]. However, it is important to note that different sectors—e.g., residential, commercial, public—face different tradeoffs and are therefore likely to respond heterogeneously to any given type of incentive [56].

PV adoption also depends heavily on the compensation paid to exported solar generation [57–59]. The most common compensation schemes worldwide have been net metering (NM) and feed-in tariffs (FIT). Under basic NM, used in many US states, each kWh of solar generation is compensated at the retail rate. Most NM schemes have the additional feature that excess solar generation credits at the end of a monthly billing cycle are rolled over to the next month, with a balancing period at the end of the year, when any accumulated energy credits are compensated at below the retail rate [60, 61]. However, many US states have adopted (or are transitioning to) a variant of NM called *net billing*, in which the compensation rate for excess generation is equal to the (significantly lower) avoided fuel cost of conventional generation [62, 63]. Conversely, with a FIT the household is paid a fixed *premium* rate per unit of exported PV generation, typically for a period of 15–25 years [64, 65]. FIT schemes for renewable energy have been widely used in Europe, Canada, Japan, and Australia, although many countries have phased out FITs for solar PV (or switched to NM compensation) in recent years, due to the overwhelming growth in solar adoption that FITs appear to have induced. **Figure 3** shows FITs paid to solar PV by year for selected countries, which peaked in most countries between 2005 and 2010. Some countries, now finding their electricity grids flooded with excess residential solar generation, are even considering *self-consumption* subsidies to encourage shifts in electricity use and limit excess solar energy export to the grid [66].

To illustrate the effectiveness of both NM and FIT incentives in expanding the market penetration of PV, one need only look at the diffusion of residential PV in markets where such policies are in place, relative to markets which have lagged behind in implementing them. In the US, California first established its NM program

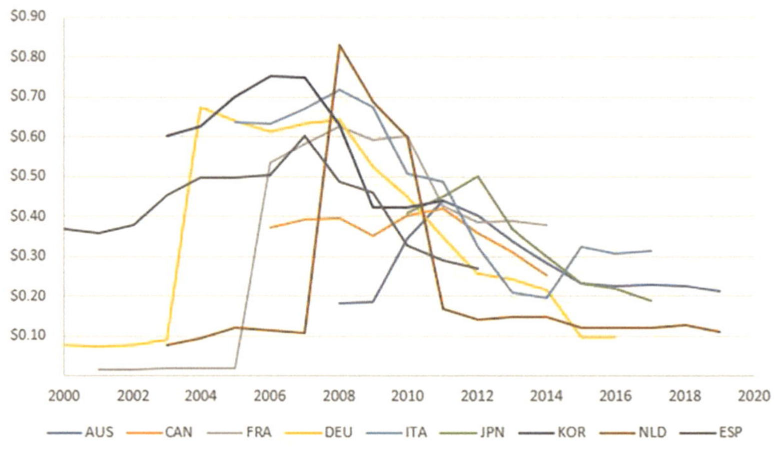

Figure 3.
Feed-in tariffs for solar PV by year for selected countries, 2000–2019 (USD/kWh). Source: OECD.

in 1995, and is now the national leader in residential PV adoption, with nearly 20% of homes having rooftop PV. By contrast, the US state of Texas has similar solar potential but no statewide NM program; as a result, less than 3% of residences have rooftop PV installed. In Europe, Germany has been a leader in solar PV due to its generous FITs, resulting in a record 3.7 million homes with rooftop PV as of 2024, and sales of BES systems are expected to increase by 26% in 2024.[1] Importantly, while most of the developed world has implemented economic support policies for rooftop PV, much of the developing world has yet to do so. Given the well known benefits of energy access to economic development [67] and the potential for technological "leapfrogging" that rooftop PV provides vis-á-vis traditional, centralized electricity grid systems [68], governments of developing countries are increasingly interested in promoting rooftop PV through NM, FITs, and other economic support schemes.

Whether widespread adoption of DERs will lead to improved welfare outcomes is still unclear, as there are many system-wide costs and benefits to account for [2, 69–71]. For example, under the existing rate design in California, interactions between retail rate structures and avoided electricity costs resulting from residential solar PV and BES adoption have led to high avoided costs and bill savings for solar consumers but substantial *cost shifting*, because in aggregate bill savings exceed avoided costs and the cost difference is borne by other ratepayers [72]. The cost-shifting problem has raised significant equity concerns, as US rooftop solar policy is generally considered to be regressive [46, 73, 74]. Higher-income households receive a disproportionate share of policy benefits, and similar to the findings of [72], the costs of solar compensation paid to high-income adopters are often passed on to relatively lower-income non-adopters. The long-term impacts of residential BES on the electricity market in Germany are studied in [75], which concludes that policy makers should focus on regulations that promote "system friendly" battery operation to mitigate negative system impacts without affecting household behavior. Others have investigated *community solar* and *community storage* as an alternative to privately-owned residential PV and BES, finding generally that properly designed local network tariffs can significantly enhance the value of both solar PV and BES availability for users [76, 77].

[1] See: "Sales of solar batteries for German homes seen rising 26% in 2024," Reuters, 14 March 2024.

5.2 Rebound effects

The expansion of residential solar PV as an affordable DER technology is not without unintended consequences. In particular, there is growing empirical evidence that households who adopt solar PV tend, on average, to increase their total electricity consumption after the PV system is installed and begins to produce energy. This phenomenon is referred to as the *solar rebound effect* (SRE). The SRE implies that the provision of solar energy from residential rooftop PV displaces grid-supplied electricity on a less than one-for-one reduction basis, meaning forecasts of CO_2 emissions reductions resulting from the proliferation of rooftop PV may be significantly overestimated.

The SRE is similar to, but distinct from, the traditionally studied rebound effect resulting from improved energy efficiency [78–83]. The rebound effect of an energy efficiency improvement for energy-using durable goods (e.g., air-conditioning systems or personal vehicles) occurs because improved energy efficiency implies a reduction in the marginal cost of energy services provided. This lower marginal cost leads to increased consumption of energy services, which may partly (or, in extreme cases, fully) offset the counterfactual energy savings implied by the efficiency improvement. A vast empirical literature has estimated rebound effects of energy efficiency improvements across a wide array of energy services, including space heating and cooling, indoor lighting, personal and commercial driving, and many others [84].

The SRE differs from the energy efficiency rebound effect precisely because it results not from an improvement in energy efficiency of the household's appliances, but from the provision of a "green" (i.e., carbon-free), zero-marginal-generation-cost source of electricity that is a perfect substitute for grid-supplied electricity as a provider of energy services. These features imply that the effect of increased supply of solar energy on a household's electricity consumption decision is both psychologically and structurally different in terms of the underlying behavioral and microeconomic foundations. Moreover, when the household is compensated for excess solar generation by means of a net-metering or feed-in tariff policy, this additional income stream may be partially allocated toward increased energy consumption [85]. The SRE is also inherently related to household income, where lower-income households, who tend to spend a larger proportion of their total monthly budget on electricity, are likely to exhibit a larger SRE [86, 87].

In its simplest form, the SRE can be modeled as an "income effect" related to net metering compensation. The following framework is provided by [88] to model the SRE on a household's total electricity consumption following installation of a solar PV system. Denote the household's optimal level of electricity consumption prior to installing a PV system as e_{pre}. The underlying microeconomic calculus of constrained utility maximization dictates that this optimal level is a function of the price of electricity, p, and the household's (monthly) income, M. Thus, we have that $e_{pre} = e_{pre}(p, M)$. Similarly, post-adoption optimal electricity consumption is $e_{post} = e_{post}(p, M, C, e_{pv})$, where C is the monthly fixed cost of the PV system and e_{pv} is the electricity generated by the PV system. Under the straightforward assumption that the monthly net metering compensation for the PV system's generation, pe_{pv}, is greater than the monthly fixed cost, it is simple to show that $e_{post} > e_{pre}$. The SRE from PV adoption is then defined as the ratio of the increase in total electricity consumption to the amount generated by the PV system:

$$SRE = \frac{e_{post} - e_{pre}}{e_{pv}}. \quad (1)$$

Consider the following numerical illustration [89]. Say a household consumes 1000 kWh per month before installing solar PV. If the household installs a PV system that generates 500 kWh per month and its monthly electricity consumption increases to 1100 kWh per month, the SRE would be $100/500 = 0.2$, or 20%. This means one-fifth of the expected carbon reduction benefits of the PV system's output are offset by an increase in electricity consumption. Estimates of the (average) magnitude of the SRE range from roughly 10–30% of the counterfactual reduction in grid-supplied electricity implied by the output of a typical residential rooftop PV system [88, 90–96].

The above example is highly simplified. The microeconomics of the SRE under a variety of net metering policy features such as rolling credits and different remuneration rates for exported solar, as well as the possibility that behavioral drivers such as "moral licensing" can also lead to a SRE are explored in [97]. Moral licensing refers to a psychological phenomenon in which performance of a good deed (like generating "green" electricity via solar PV) leads an individual to feel liberated to engage in socially damaging actions (like increased energy consumption) and has been studied as a likely driver of energy efficiency rebound effects [98, 99]. Any number of other behavioral phenomena may also be at work to ameliorate or exacerbate the SRE (and rebound effects, generally), including "warm glow" effects, bounded rationality, mental accounting, or moral cleansing [100, 101]. More research is needed to uncover how much of the SRE can be attributed to a simple income effect of net metering (or other) compensation and how much is explained by behavioral effects.

Additionally, much research is still needed to understand how the SRE affects the overall benefit-cost ratio of subsidizing distributed energy generation through residential rooftop solar PV. The net effect on total economic welfare may be either positive or negative, depending on factors such as how clean or dirty the underlying electricity system is, local climate/weather conditions, and population density [97]. If the net effect of the SRE on total economic welfare is positive, this would further support the use of public subsidies to incentivize adoption of residential rooftop PV systems. Conversely, if the SRE leads to a net reduction in economic welfare, this may justify shifting subsidy dollars toward utility-scale solar, which, as [89] argues, is unlikely to result in a rebound effect.

DERs like solar PV and small-scale BES may also lead to other types of rebound effects. Indirect rebound effects may arise if cost savings from solar energy generation are reallocated toward increased consumption of other fuels (e.g., natural gas). Some analyses have even modeled long-run, macroeconomic rebound effects resulting from reduced energy costs that may lead to increased economic activity and, as a result, aggregate energy consumption [81, 102–105]. Such effects of widespread diffusion of DERs have yet to be documented in the literature, but may arise as market penetration continues to increase.

6. Conclusions

The market for distributed energy technologies like rooftop solar PV and BES continues to expand worldwide. Manufacturing and installation costs have declined precipitously and will continue to do so for the foreseeable future. Consumer preferences are shifting in favor of decentralized, clean, and carbon-free technologies.

Governments—concerned both with mitigating climate change and improving grid resiliency—are actively supporting the development and adoption of these technologies through a variety of economic support schemes.

However, the increasingly widespread diffusion of DERs presents significant challenges for electricity system operators in terms of integrating these resources into the broader energy mix. The goal of this chapter has been to review the economics of this integration process, focusing on several key areas of ongoing economic and policy research. A prominent theme that has emerged is that the entirety of the electricity supply chain—from generation to transmission to retail distribution, as well as the regulatory bodies responsible for governing electricity markets at each level—will be required to make significant changes if it is to adapt to the emerging decentralized clean energy paradigm. Integrating these resources efficiently will require rethinking many of the regulatory, operational, financial, and market-based approaches to managing electricity systems.

The economic and policy landscape surrounding DER market penetration—and the electricity sector, generally—is changing rapidly. At the same time, different scenarios of DER adoption and technology costs will affect the relevant economic and policy tradeoffs. Will the cost of production for BES technologies decline as dramatically as they have for solar PV? How will technological innovations and cost reductions evolve over the next 10–15 years? As has been conveyed in this chapter, the answers to these and other key questions are being actively researched by energy economists and energy policy analysts and will largely determine the benefit-cost ratios of policies designed to support research and development of DER technologies, limit the adverse effects of intermittency on the electricity system, alter consumer behavior, and ultimately increase market penetration of DERs. These tradeoffs must be managed thoughtfully if DERs are to achieve decentralization and deep decarbonization of the electricity sector in a way that is both efficient and sustainable.

Thanks

The author thanks Maghfira Ramadhani for research assistance.

Abbreviations

BES	battery electric storage
CCNG	combined-cycle natural gas
DER	distributed energy resource
FIT	feed-in tariff
FRP	flexible ramping product
IERP	integrated electricity resource planning
IPP	independent power producer
ISO	independent system operator
kWh	kilowatt hour
LCOE	levelized cost of energy
MOD	merit-order dispatch
MWh	megawatt hour
MMP	missing-money problem
NM	net metering
PV	photovoltaics

RA	resource adequacy
SRE	solar rebound effect
VOLL	value of lost load

Author details

Matthew E. Oliver
School of Economics, Georgia Institute of Technology, Atlanta, USA

*Address all correspondence to: matthew.oliver@econ.gatech.edu

References

[1] US Energy Information Administration. International Energy Outlook 2023. Washington, DC, USA: US Department of Energy; 2023

[2] Agnew S, Dargusch P. Effect of residential solar and storage on centralized electricity supply systems. Nature Climate Change. 2015;**5**(4): 315-318

[3] Cretì A, Fontini F. Economics of Electricity: Markets, Competition and Rules. UK: Cambridge University Press; 2019

[4] Clò S, Cataldi A, Zoppoli P. The merit-order effect in the Italian power market: The impact of solar and wind generation on national wholesale electricity prices. Energy Policy. 2015;**77**: 79-88

[5] Cludius J, Hermann H, Matthes FC, Graichen V. The merit order effect of wind and photovoltaic electricity generation in Germany 2008–2016: Estimation and distributional implications. Energy Economics. 2014; **44**:302-313

[6] Johnson EP, Oliver ME. Renewable generation capacity and wholesale electricity price variance. The Energy Journal. 2019;**40**(5):143-168

[7] Genoese F, Genoese M, Wietschel M. Occurrence of negative prices on the German spot market for electricity and their influence on balancing power markets. In: 2010 7th International Conference on the European Energy Market. Madrid, Spain: IEEE; 2010. pp. 1-6

[8] Seel J, Millstein D, Mills A, Bolinger M, Wiser R. Plentiful electricity turns wholesale prices negative. Advances in Applied Energy. 2021;**4**:100073

[9] Skea J, Anderson D, Green T, Gross R, Heptonstall P, Leach M. Intermittent renewable generation and the cost of maintaining power system reliability. IET Generation, Transmission & Distribution. 2008;**2**(1):82-89

[10] Liebensteiner M, Wrienz M. Do intermittent renewables threaten the electricity supply security? Energy Economics. 2020;**87**:104499

[11] Godwin V, Oliver ME. Wind intermittency and supply-demand imbalance: Evidence from US regional power markets. USAEE Working Paper 23-588; 2024. 10.2139/ssrn.4739240

[12] Godoy-González D, Gil E, Gutiérrez-Alcaraz G. Ramping ancillary service for cost-based electricity markets with high penetration of variable renewable energy. Energy Economics. 2020;**85**:104556

[13] Gowrisankaran G, Reynolds SS, Samano M. Intermittency and the value of renewable energy. Journal of Political Economy. 2016;**124**(4):1187-1234

[14] Ambec S, Crampes C. Decarbonizing electricity generation with intermittent sources of energy. Journal of the Association of Environmental and Resource Economists. 2019;**6**(6): 1105-1134

[15] Helm C, Mier M. Steering the energy transition in a world of intermittent electricity supply: Optimal subsidies and taxes for renewables and storage. Journal of Environmental Economics and Management. 2021;**109**:102497

[16] Varghese S, Sioshansi R. The price is right? How pricing and incentive mechanisms in California incentivize building distributed hybrid solar and

energy-storage systems. Energy Policy. 2020;**138**:111242

[17] Giulietti M, Grossi L, Baute ET, Waterson M. Analyzing the potential economic value of energy storage. The Energy Journal. 2018;**39**(1 suppl): 101-122

[18] Geske J, Green R. Optimal storage, investment and management under uncertainty: It is costly to avoid outages! The Energy Journal. 2020;**41**(2):1-28

[19] Schmalensee R. Competitive energy storage and the duck curve. The Energy Journal. 2022;**43**(2):1-16

[20] Carson RT, Novan K. The private and social economics of bulk electricity storage. Journal of Environmental Economics and Management. 2013; **66**(3):404-423

[21] Stoft S. Power System Economics: Designing Markets for Electricity. New Jersey: IEEE Press; 2002

[22] Joskow PL. Challenges for wholesale electricity markets with intermittent renewable generation at scale: The US experience. Oxford Review of Economic Policy. 2019;**35**(2):291-331

[23] Kahrl F, Mills A, Lavin L, Ryan N, Olsen A, Schwartz L. The Future of Electricity Resource Planning. Berkeley, CA (United States): Lawrence Berkeley National Lab (LBNL); 2016

[24] Cretì A, Fabra N. Supply security and short-run capacity markets for electricity. Energy Economics. 2007; **29**(2):259-276

[25] Cramton P, Ockenfels A, Stoft S. Capacity market fundamentals. Economics of Energy & Environmental Policy. 2013;**2**(2):27-46

[26] Holmberg P, Ritz RA. Optimal capacity mechanisms for competitive electricity markets. The Energy Journal. 2021;**42**(1 suppl):1-34

[27] Tanaka M. Real-time pricing with ramping costs: A new approach to managing a steep change in electricity demand. Energy Policy. 2006;**34**(18): 3634-3643

[28] Khoshjahan M, Fotuhi-Firuzabad M, Moeini-Aghtaie M, Dehghanian P. Enhancing electricity market flexibility by deploying ancillary services for flexible ramping product procurement. Electric Power Systems Research. 2021;**191**:106878

[29] Schmidt O, Hawkes A, Gambhir A, Staffell I. The future cost of electrical energy storage based on experience rates. Nature Energy. 2017;**2**(8):1-8

[30] Junge C, Mallapragada D, Schmalensee R. Energy storage investment and operation in efficient electric power systems. The Energy Journal. 2022;**43**(6):1-24

[31] Siddiqui AS, Sioshansi R, Conejo AJ. Merchant storage investment in a restructured electricity industry. The Energy Journal. 2019;**40**(4):129-164

[32] Joskow PL. Facilitating transmission expansion to support efficient decarbonization of the electricity sector. Economics of Energy & Environmental Policy. 2022;**10**(2):57-91

[33] Davis LW, Hausman C, Rose NL. Transmission impossible? Prospects for decarbonizing the US grid. The Journal of Economic Perspectives. 2023;**37**(4): 155-180

[34] Li HX, Horan P, Luther MB, Ahmed TM. Informed decision making of battery storage for solar-PV homes

using smart meter data. Energy and Buildings. 2019;**198**:491-502

[35] Lo CH, Ansari N. Alleviating solar energy congestion in the distribution grid via smart metering communications. IEEE Transactions on Parallel and Distributed Systems. 2012; **23**(9):1607-1620

[36] Castaneda M, Franco CJ, Dyner I. Evaluating the effect of technology transformation on the electricity utility industry. Renewable and Sustainable Energy Reviews. 2017;**80**: 341-351

[37] Athawale R, Felder FA. Electric utility death spiral: Revisited in the context of tariff design. The Electricity Journal. 2022;**35**(1):107062

[38] Castaneda M, Jimenez M, Zapata S, Franco CJ, Dyner I. Myths and facts of the utility death spiral. Energy Policy. 2017;**110**:105-116

[39] Laws ND, Epps BP, Peterson SO, Laser MS, Wanjiru GK. On the utility death spiral and the impact of utility rate structures on the adoption of residential solar photovoltaics and energy storage. Applied Energy. 2017; **185**:627-641

[40] Muaafa M, Adjali I, Bean P, Fuentes R, Kimbrough SO, Murphy FH. Can adoption of rooftop solar panels trigger a utility death spiral? A tale of two US cities. Energy Research & Social Science. 2017;**34**:154-162

[41] Khalilpour R, Vassallo A. Leaving the grid: An ambition or a real choice? Energy Policy. 2015;**82**:207-221

[42] Costello KW, Hemphill RC. Electric utilities 'death spiral': Hyperbole or reality? The Electricity Journal. 2014; **27**(10):7-26

[43] Best R, Li H, Trück S, Truong C. Actual uptake of home batteries: The key roles of capital and policy. Energy Policy. 2021;**151**:112186

[44] Günther C, Schill WP, Zerrahn A. Prosumage of solar electricity: Tariff design, capacity investments, and power sector effects. Energy Policy. 2021;**152**: 112168

[45] Brown DP. Socioeconomic and demographic disparities in residential battery storage adoption: Evidence from California. Energy Policy. 2022;**164**: 112877

[46] Borenstein S. Private net benefits of residential solar PV: The role of electricity tariffs, tax incentives, and rebates. Journal of the Association of Environmental and Resource Economists. 2017;**4**(S1):S85-S122

[47] Feger F, Pavanini N, Radulescu D. Welfare and redistribution in residential electricity markets with solar power. The Review of Economic Studies. 2022; **89**(6):3267-3302

[48] Andreolli F, D'Alpaos C, Moretto M. Valuing investments in domestic PV-battery systems under uncertainty. Energy Economics. 2022;**106**:105721

[49] Hassi B, Reyes T, Sauma E. A compound real option approach for determining the optimal investment path for RPV-storage systems. The Energy Journal. 2022;**43**(3):83-104

[50] Gillingham K, Deng H, Wiser R, Darghouth N, Nemet G, Barbose G, et al. Deconstructing solar photovoltaic pricing: The role of market structure, technology, and policy. The Energy Journal. 2016;**37**(3):231-250

[51] Sarzynski A, Larrieu J, Shrimali G. The impact of state financial incentives

on market deployment of solar technology. Energy Policy. 2012;**46**: 550-557

[52] Hughes JE, Podolefsky M. Getting green with solar subsidies: Evidence from the California solar initiative. Journal of the Association of Environmental and Resource Economists. 2015;**2**(2):235-275

[53] Crago CL, Chernyakhovskiy I. Are policy incentives for solar power effective? Evidence from residential installations in the northeast. Journal of Environmental Economics and Management. 2017;**81**:132-151

[54] Matisoff DC, Johnson EP. The comparative effectiveness of residential solar incentives. Energy Policy. 2017; **108**:44-54

[55] Sun B, Sankar A. The changing effectiveness of financial incentives: Theory and evidence from residential solar rebate programs in California. Energy Policy. 2022;**162**:112804

[56] Rydehell H, Lantz B, Mignon I, Lindahl J. The impact of solar PV subsidies on investment over time-the case of Sweden. Energy Economics. 2024;**133**:107552

[57] Darghouth NR, Barbose G, Wiser R. The impact of rate design and net metering on the bill savings from distributed PV for residential customers in California. Energy Policy. 2011;**39**: 5243-5253

[58] Brown DP, Sappington DEM. Designing compensation for distributed solar generation: Is net metering ever optimal? The Energy Journal. 2017;**38**(3): 1-32

[59] Gautier A, Jacqmin J. PV adoption: The role of distribution tariffs under net metering. Journal of Regulatory Economics. 2020;**57**:53-73

[60] Hughes L, Bell J. Compensating customer-generators: A taxonomy describing methods of compensating customer-generators for electricity supplied to the grid. Energy Policy. 2006;**34**:1532-1539

[61] Eid C, Guillén JR, Marín PF, Hakvoort R. The economic effect of electricity net-metering with solar PV: Consequences for network cost recovery, cross subsidies and policy objectives. Energy Policy. 2014;**75**: 244-254

[62] Comello S, Reichelstein S. Cost competitiveness of residential solar PV: The impact of net metering restrictions. Renewable and Sustainable Energy Reviews. 2017;**75**:46-57

[63] Schelly C, Louie EP, Pearce JM. Examining interconnection and net metering policy for distributed generation in the United States. Renewable Energy Focus. 2017;**22**:10-19

[64] Couture T, Gagnon Y. An analysis of feed-in tariff remuneration models: Implications for renewable energy investment. Energy Policy. 2010;**38**: 955-965

[65] Mendonça M, Jacobs D, Sovacool B. Powering the Green Economy: The Feed-in Tariff Handbook. London: Earthscan; 2010

[66] Colasante A, D'Adamo I, Morone P. What drives the solar energy transition? The effect of policies, incentives and behavior in a cross-country comparison. Energy Research & Social Science. 2022;**85**:102405

[67] Toman MA, Jemelkova B. Energy and economic development: An

assessment of the state of knowledge. The Energy Journal. 2003;**24**(4):93-112

[68] Fetter TR. Energy transitions and technology change: "Leapfrogging" reconsidered. Resource and Energy Economics. 2022;**70**:101327

[69] Schill WP, Zerrahn A, Kunz F. Prosumage of solar electricity: Pros, cons, and the system perspective. Economics of Energy & Environmental Policy. 2017;**6**(1):7-32

[70] Ali GG, El-adaway IH, Sims C, Holladay JS, Chen CF. Policies and incentives for promoting distributed solar generation: Impact on electric power infrastructure. Journal of Infrastructure Systems. 2022;**28**(4): 04022034

[71] Simshauser P, Wild P. Rooftop solar PV, coal plant inflexibility and the minimum load problem. CAEEPR Working Paper 2023-04; 2023

[72] Boampong R, Brown DP. On the benefits of behind-the-meter rooftop solar and energy storage: The importance of retail rate design. Energy Economics. 2020;**86**:104682

[73] Borenstein S, Davis LW. The distributional effects of US clean energy tax credits. Tax Policy and the Economy. 2016;**30**(1):191-234

[74] Dauwalter TE, Harris RI. Distributional benefits of rooftop solar capacity. Journal of the Association of Environmental and Resource Economists. 2023;**10**(2):487-523

[75] Fett D, Fraunholz C, Keles D. Diffusion and system impact of residential battery storage under different regulatory settings. Energy Policy. 2021;**158**:112543

[76] Keck F, Lenzen M. Drivers and benefits of shared demand-side battery storage–an Australian case study. Energy Policy. 2021;**149**:112005

[77] Sturmberg B, Shaw M, Mediwaththe C, Ransan-Cooper H, Weise B, Thomas M, et al. A mutually beneficial approach to electricity network pricing in the presence of large amounts of solar power and community-scale energy storage. Energy Policy. 2021;**159**:112599

[78] Khazzoom JD. Economic implications of mandated efficiency in standards for household appliances. The Energy Journal. 1980;**1**(4):21-40

[79] Berkhout PHG, Muskens JC, Velthuijsen JW. Defining the rebound effect. Energy Policy. 2000;**28**(6–7): 425-432

[80] Sorrell S, Dimitropoulos J. The rebound effect: Microeconomic definitions, limitations and extensions. Ecological Economics. 2008;**65**(3): 636-649

[81] Azevedo IML. Consumer end-use energy efficiency and rebound effects. Annual Review of Environment and Resources. 2014;**39**:393-418

[82] Borenstein S. A microeconomic framework for evaluating energy efficiency rebound and some implications. The Energy Journal. 2015; **36**(1):1-21

[83] Chan NW, Gillingham K. The microeconomic theory of the rebound effect and its welfare implications. Journal of the Association of Environmental and Resource Economists. 2015;**2**(1):133-159

[84] Sorrell S, Dimitropoulos J, Sommerville M. Empirical estimates of

the direct rebound effect: A review. Energy Policy. 2009;**37**(4):1356-1371

[85] La Nauze A. Power from the people: Rooftop solar and a downward-sloping supply of electricity. Journal of the Association of Environmental and Resource Economists. 2019;**6**(6): 1135-1168

[86] Oliver ME, Moreno-Cruz J, Beppler RC. Microeconomics of the rebound effect for residential solar photovoltaic systems. CESifo Working Paper No 7635; 2019

[87] Toroghi SH, Oliver ME. Framework for estimation of the direct rebound effect for residential photovoltaic systems. Applied Energy. 2019;**251**: 113391

[88] Beppler RC, Matisoff DC, Oliver ME. Electricity consumption changes following solar adoption: Testing for a solar rebound. Economic Inquiry. 2023;**61**(1):58-81

[89] Oliver ME. Tipping the scale: Why utility-scale solar avoids a solar rebound and what it means for US solar policy. The Electricity Journal. 2023;**36**(4): 107266

[90] Havas L, Ballweg J, Penna C, Race D. Power to change: Analysis of household participation in a renewable energy and energy efficiency programme in Central Australia. Energy Policy. 2015;**87**:325-333

[91] Deng G, Newton P. Assessing the impact of solar PV on domestic electricity consumption: Exploring the prospect of rebound effects. Energy Policy. 2017;**110**:313-324

[92] Spiller E, Sopher P, Martin N, Mirzatuny M, Zhang X. The environmental impacts of green technologies in TX. Energy Economics. 2017;**68**:199-214

[93] Qiu Y, Kahn ME, Xing B. Quantifying the rebound effects of residential solar panel adoption. Journal of Environmental Economics and Management. 2019;**96**:310-341

[94] Boccard N, Gautier A. Solar rebound: The unintended consequences of subsidies. Energy Economics. 2021; **100**:105334

[95] Frondel M, Kaestner K, Sommer S, Vance C. Photovoltaics and the solar rebound: Evidence from Germany. Land Economics. 2022;**98**(4):265-282

[96] Aydın E, Brounen D, Ergün A. The rebound effect of solar panel adoption: Evidence from Dutch households. Energy Economics. 2023;**120**:106645

[97] Oliver ME, Moreno-Cruz J, Gillingham K. Microeconomics of the solar rebound under net metering. USAEE Working Paper No 23-588; 2023

[98] Dütschke E, Frondel M, Schleich J, Vance C. Moral licensing: Another source of rebound? Frontiers in Energy Research. 2018;**6**:38

[99] Peters A, Dütschke E. Exploring rebound effects from a psychological perspective. In: Rethinking Climate and Energy Policies. Switzerland: Springer; 2016. pp. 89-105

[100] Dorner Z. A behavioral rebound effect. Journal of Environmental Economics and Management. 2019;**98**: 102257

[101] Exadaktylos F, van den Bergh J. Energy-related behaviour and rebound when rationality, self-interest, and willpower are limited. Nature Energy. 2021;**6**:1104-1113

[102] Bentzen J. Estimating the rebound effect in US manufacturing energy consumption. Energy Economics. 2004; **26**(1):123-134

[103] Barker T, Ekins P, Foxon T. The macro-economic rebound effect and the UK economy. Energy Policy. 2007; **35**(10):4935-4946

[104] Yu X, Moreno-Cruz J, Crittenden JC. Regional energy rebound effect: The impact of economy-wide and sector level energy efficiency improvement in Georgia, USA. Energy Policy. 2015;**87**:250-259

[105] Blackburn C, Moreno-Cruz J. Energy efficiency in general equilibrium with input-output linkages. Journal of Environmental Economics and Management. 2021;**110**:102542

Chapter 5

Power Quality Monitoring and Equipment in Distribution Microgrids: Insights from a Case Study

Laurentiu Constantin Lipan and George Alexandru Florea

Abstract

This chapter aims to contribute to the integration of classical energy sources, aiming for zero carbon emissions and renewable photovoltaic energy in isolated sections of the electricity grid. By conducting a broad contextual analysis encompassing economic, environmental, and policy considerations, the chapter seeks to enhance the positive impact of these integrations, addressing the multifaceted challenges and opportunities in sustainable electrical system operations. A detailed case study will illustrate specific system operations, positioning this example within broader trends in smart network development and highlighting the challenges of integrating renewable and classical energy sources in alignment with sustainable development objectives. Beyond technical performance, the chapter will assess the economic viability and environmental impact of combining diesel generators with photovoltaic systems, offering a comprehensive picture of system sustainability. A comparative analysis of the Master-Slave system configuration with other control strategies will identify advantages and areas for improvement, providing valuable insights for future implementations. Additionally, the chapter will discuss the fit of such systems within current policy frameworks and potential regulatory obstacles, offering guidance to policymakers and stakeholders to facilitate the adoption of similar sustainable energy solutions.

Keywords: diesel group, recloser, accumulators, performance standard, programmable automatic

1. Introduction

In this chapter, we start from the premise that the operation of generator groups is quite frequent or even permanent, as the analyzed industrial area currently lacks the capability to operate at the required power and load. Thus, most of the time, Master-Slave generator sets will be used to power internal systems, automations, and accumulators from a micro-photovoltaic system. This system typically operates in isolation until the completion of the entire electrical power supply system, such as in oil fields during drilling, soil tests, or when anomalies necessitate the construction of electrical stations, mobile power supply points, or site organizations (which may last over 2 to 5 years) in certain areas [1], especially during pandemics or wars. A recent concern within industrial users is the quality of electricity, including its planning, monitoring, and verification of compliance with specific technical standards regarding compatibility levels, harmonics, unbalance, flicker, and other power quality elements. It is crucial for every supplier to analyze the causes and create a statistical basis for correlating disturbances with their effects, including phenomena like overvoltages, overloads, and rapid wear, when the designed technical and economic parameters are not met.

The primary role of the electricity network and its operation is to provide electricity to users in accordance with energy regulations and standards that govern the design, construction, operation, maintenance, distribution, and supply of electricity [1–3].

To fulfill this role, the microgrid must meet the following technical and financial requirements:

- Safety in operation – constitutes a series of measures taken in the design and operation of electrical installations to meet emerging requirements as quickly as possible, without interruptions or downtime (as small as possible). Compliance with this requirement requires the following reliability values to be met for a specified period of time (usually 1 year):
    - the average number and maximum number of interruptions;
    - average duration and maximum duration of the interruption;
- The quality of electricity – it is evaluated by means of the quality indicators of specific electrical sizes, voltage, frequency and power supply service respectively by power disconnection times. Quality indicators have achieved the points of separation between the electrical installations from the supplier and the user who make up the burden:
    - the supplier for: frequency, amplitude and voltage gaps, temporary and transient overvoltage's, duration of interruptions;
    - the consumer to: harmonic and interharmonic, voltage deviations, unbalance.

The permissible values for most of the quality indicators are determined by energy standards and requirements.

- Economics – acknowledges that in order to maximum efficiency in terms of revenue-expenditure, technical and economic standards and also procedures for the optimization and operation of the electricity grid must be applied. The application of these technical and economic standards and a specific process, allows the solution of the following problems:
    - selection of solutions for fulfilling energy objectives;
    - specification of appropriate regimes from an economic perspective for the operation of electrical equipment (electric lines, power transformers etc.);

- o determination with Figures or modification of electricity networks of transmission and distribution so that we obtain economic power distribution.

- Environmental impact – aims at reducing or eliminating the environmental impact caused by energy targets:

  - o the ecological impact, the causes of soil, water, air pollution, activity on different categories of land, destruction of vegetation, fauna, etc.

  - o electromagnetic effects on people or household appliances due to the location of the network in a residential area or due to side effects (voltage fluctuations).

The visual impact is caused by the locations of the electricity grid (electric lines, electric-stations, transformation stations, etc.) in the environment [1, 2].

Micro electrical station thus formed, according to **Figure 1**, presents, in addition to the usual income, supply, measurement cells, longitudinal torque, etc., also a series of connection cells through which the injection of electrical power supplied by the corresponding GENSET groups can be achieved (G1–G2 groups for a section/branch and G3, G4,..., G8 groups for the rest). The longitudinal couplings are commanded by the ATS (Auto Switching of the Reserve) and also by a modern RECLOSER.

Recloser is a useful tool, increasingly found on industrial platforms [1], in the industrial and tertiary environment, both on the part of ensuring safety in operation, ensuring the reserve of electricity, increasing the speed of intervention in terms of switching to the power supply of a user in case of a failure, isolating a certain area in the case of defect, as well as on the side, monitoring the quality of power supply [2], but also other electrical parameters related to electromagnetic compatibility, electromagnetic field, the resistance of the ground outlet and other electric but also non-electric sizes.

Microgrids are increasingly recognized as a pivotal component in modern energy systems, offering enhanced resilience, reliability, and efficiency in power distribution. These localized networks, which can operate independently or in conjunction with the main grid, provide significant benefits, including improved energy security,

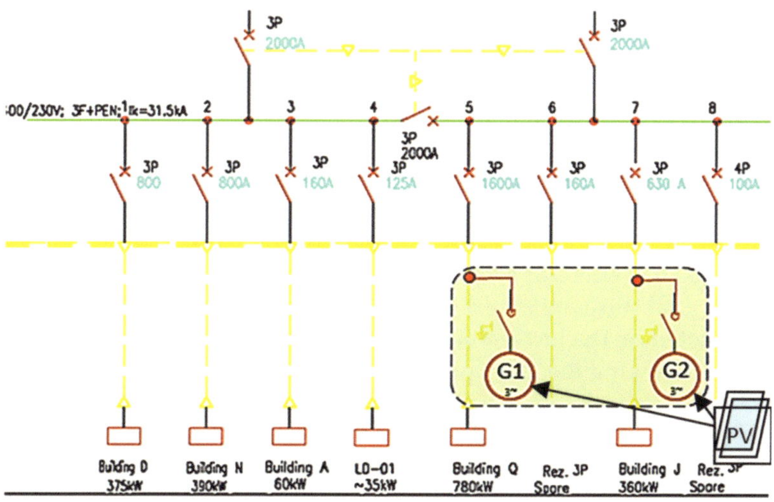

Figure 1.
General electricity supply scheme of the user.

integration of renewable energy sources, and tailored energy management. However, the effective operation of microgrids is contingent upon maintaining high power quality standards. Power quality issues in microgrids—such as voltage fluctuations, harmonic distortions, and frequency imbalances—can adversely affect the performance of sensitive electronic equipment, compromise system stability, and reduce overall efficiency. Addressing these power quality challenges is crucial for ensuring the reliable and efficient operation of microgrids, as well as for achieving their potential in supporting sustainable energy transitions and enhancing grid resilience.

2. Analysis of the quality of the power supply of the three-phase power supply circuit of the group diesel generator G1 (560KVA, 400 V) and of the group diesel generator G2 (715kVA, 400 V)

To carry out the measurements, in such a case, when, due to technical problems, part of the analyzed user's sub-consumers are supplied with electricity in a way considered islanded (on a portion of the electrical network), in which they intervene in turn in a Master-Slave mode, after maintaining for a short period of time the voltage level and respectively the load in the accumulator batteries, the two gensets G1 and G2 intervene in turn (depending on needs and settings).

In the general supply/discharge electrical panel of each generator group separately, on the three bars/phases, at the exit from the general automatic switch/contactor for connecting the diesel group, two analyzers, monitoring equipment type ION7600/Ion7650 (one for each group separately) [3].

Given the measurement precision class of the monitoring devices used (Class A according to IEC/CEI standards 61,000, IEC 61000-3-2, IEC 61000-6-2, IEC 61000-3-4, IEC 61000-3-6), the analysis method is a direct one based on on-site measurements conducted over a period deemed representative, using a specific case study or pilot. This case, which is a simplified industry example without special automation or reclosers on the low-voltage side, can be generalized and applied to other industrial areas and schemes. It can effectively be implemented in various low-voltage contexts, from hospitals and small industrial zones to neighborhoods, residential complexes, and airport terminals, particularly in the event of prolonged blackouts leading to transformer damage. Transformer deterioration necessitates extended interventions, making the rapid deployment of such systems crucial for restoring power to users who are left without electricity.

Regarding the operation of these generator groups (GENSETS) G1 and G2 in an automated Master-Slave system, their coordination is carried out through a modern system of the RECLOSER & ATS type on the low-voltage side, at an operation of the analyzed user in isolated mode, independent of the rest of the power system (**Figure 1**) [1, 2]. The generator sets G1 and G1 are the pair of stand-by groups that assist a series of vital sub-consumers of particular importance to the targeted industrial user, connected on the low-voltage side to the user's electrical network. These groups have the following installed powers:

1. G1 – 560kVA at 400 V (three-phase), with the electrical part governed by a modern controller connected to a Recloser (including the role of ATS);

2. G2 – 715kVA at 400 V (three-phase), with the electrical part governed by a modern controller at a Recloser (including the role of ATS).

The selection of the assisted (electricity supply) bar and the respective genset (connected to that bar) is made through a specialized electrical switching and selection panel. The role of this electrical panel, which exists in the internal electrical distribution scheme with the RECLOSER and the ATSs belonging to the stand-by groups G1 and G2, is to assist the networks of critical sub-consumers, with the necessary gensets (generators).

The system for ensuring the continuity of the user/consumers' electricity supply, TCS – Switchboard and Selection, coordinates the intervention of the accumulator batteries and the intervention groups, being made up of:

- related electrical installation;
- interconnected SCADA type system that ensures, through one of its components, the intervention of the accumulator batteries, but also the operation of the two gensets (G1 and G2) in "Master-Slave" mode; that is, upon reaching a preset load capacity of a certain number of percent of its maximum capacity, the automatic start of the "slave" group G2 is determined;
- modernized RUT type rectifiers 220Vcc/100A that provide the buffer voltage for the stationary battery of 220 V accumulators;
- Specialized ATS & Recloser.

When the group identified and set as MASTER reaches maximum capacity, it coordinates the groups set as SLAVE to start as follows: MASTER at maximum capacity, turns on SLAVE 1, which, if it also reaches maximum capacity, is commanded, also by the MASTER group to start the SLAVE 2 group); we must also take into account the mandatory connection in the SCADA-type system.

The monitoring system is complex; it ensures the measurement and monitoring of the electric line currents (in the connection boxes, electric stations, and electrical substations), the state of the surge arresters, the existing electrical filters, the outside temperature, the temperatures of the transformer modules, including a complete analysis of the consumption of energy and power circulation on branches/loops, practically through a complete electrical network management system [1, 4, 5].

The SCADA system, according to the IEC 61850 standards, ensures the conditions of interoperability between the existing IEDs and the new connections (even with devices from different manufacturers) thus the semantics of the monitored, analyzed, transmitted, inter-exchanged data is already completely standardized, as well as communication procedures (fully functional bidirectional interoperability). Any recipient must understand all data transmitted by any sender without human intervention. That is, the protection relays practically communicate with each other, as well as with the type II generation SCADA system, with implemented IEC61850 protocol. There are a number of theoretical and constructive technical elements presented in a separate chapter for Recloser in the present exposition [3, 6].

Thus, in the G1 – Master group (**Figure 1a**) analyzed in this chapter, an ION™ 7600 class A analyzer was installed in the time interval 01/28/2022 at 07:02:52 AM to 28/01/2022 at 07:09:28 AM to monitor various electrical quantities as will be described in the following. The use of ION™ 7600 aimed to perform an energy analysis of these quantities for the technological process at the general connection supply station located at the supply point within the company's premises.

Its effective operation (intervention of the G1 group) took place on 01/28/2022 at 07:03:06 AM (when a 9.1A load appears on phase 3, the others being zero) until 01/28/2022 at 7:00 AM:09:28 AM (when the load on G1 is disconnected, G1 thus having zero load and the voltage remaining constant at the terminals with the values of 409.6 V / 411.1 V/410.3 V until monitoring is stopped).

Thus, the analysis of the data recorded from 01/28/2022 at 07:02:52 AM to 28/01/2022 at 07:09:28 AM for supplying the receivers within the company, subject to analysis, allows highlighting the electrical characteristics of the electricity supply system. To carry out the electric power study within the company, the measurements performed on the low-voltage (LV) bars during the above-mentioned period were used, with the recording of the average values of the electrical quantities at an interval of 1 second (1 s).

The voltage at the power supply bars of the receivers shows variations (236...241) Voltage per phase/ (409...418) Voltage between phases for TGD (General Distribution Table – Electric panel) in **Figure 2a**, which must fall within the limits of 230 ± 10% / 400 ± 10% V (207...253/360...440 V). Rapid variations in voltage values, caused by specific processes within the enterprise, result in recorded voltage fluctuations accompanied by the flicker effect. The three phases are relatively symmetrically loaded. It is observed that the supply voltage values during the measurement period remained within the normal (normative) limits. During 01/28/2022 at 07:04:5 PM to 28/01/2022 08:12:25 AM, at the terminals of the generator group G2 (**Figure 1c**) high-performance equipment for monitoring and measuring some elements of the quality of electricity type ION™ 7600 class A was mounted; its effective operation (intervention of the G2 group) took place during 01/28/2022 at 07.04:17 AM (when a load of Arms Line1 = 77.7A / Arms Line2 = 77.9A / Arms Line3 = 80.9A/Arms Neutral = 14A appears, the others being zero) until 28/01/2022 at 07:09:28 AM (when the load on G1-Master is disconnected, G2 thus having zero load on all three phases and the voltage remaining constant at the terminals with the values of 404.6 V / 404.4 V/404.6 V until the monitoring is stopped).

Thus, the analysis of the data recorded from 01/28/2022 at 07:02:52 AM to 28/01/2022 at 07:09:28 AM, for supplying the receivers within the company, subject to analysis, allows highlighting the electrical characteristics of the electricity supply system.

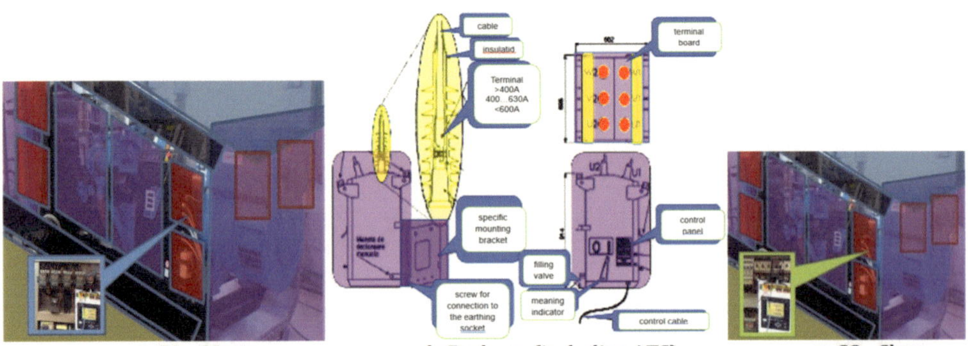

a. G1 – Master b. Recloser (including ATS) c. G2 - Slave

Figure 2.
Schematic presentation of the connection of Generator Groups G1 and G2 in the Master System through a Recloser. a. The Diesel G1 generator set and the method of mounting the ION 7600 Class A analyzer at the group's terminals; b. Recloser and automation; c. Diesel G2 genset; the method of mounting the ION 7600 Class A analyzer at the terminals of the generator group [1, 2].

The voltage at the power bars of the receivers powered and from G2-Slave shows variations (221.8...223.1) Voltage per phase/(383.2...385) Voltage between phases for TGD (**Figure 3a** and **b**), which must fall within the limits of 230 ± 10% /400 ± 10% V (207···257/360···440 V). Rapid variations in voltage values caused by specific processes within the enterprise lead to the recording of voltage fluctuations accompanied by the flicker effect. The three phases are loaded relatively symmetrically. It is noted that the supply voltage values during the measurement period remained within the normal (normative) limits. The voltages from the supply bars have a waveform close to a sine wave, characterized by relatively low total harmonic distortion voltage (THD-U) (**Figure 4**), which falls within the permissible limits for low-voltage supply bars (THD-U_{admis} is below 8%). The total voltage distortion factor is relatively low, ranging between 1.2 and 3.6% [5–7]. The variation of the voltage unbalance factor during the recording period is shown in **Figure 5** and the variation of electric current unbalance factor during the recording period is shown in **Figure 6**. The negative unbalance factor was calculated using the following equation:

$$k_s = \frac{I^-}{I^+} \tag{1}$$

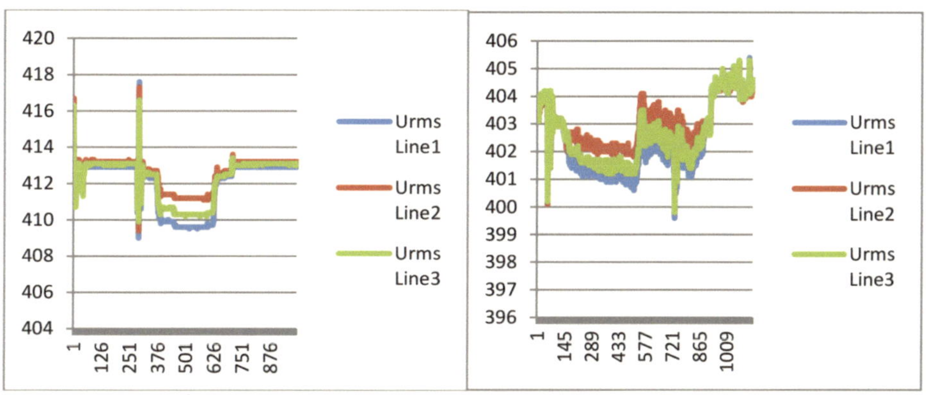

Measurement period – Monitoring time
a. G1 – Master
I. Urms [V]

Measurement period – Monitoring time
b. G2 – Slave

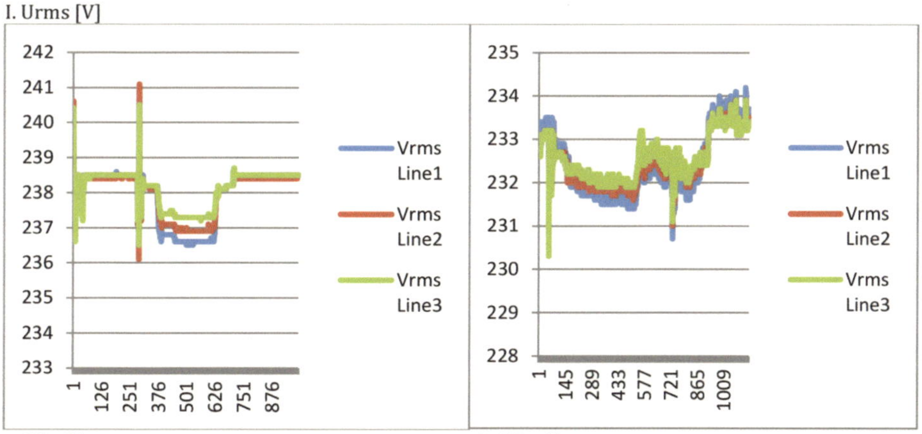

a. G1 – Master
II. Vrms [V]

b. G2 – Slave

Figure 3.
Voltage curve at the supply bus for TGD: Iw – voltages between phases, II - voltages per phase.

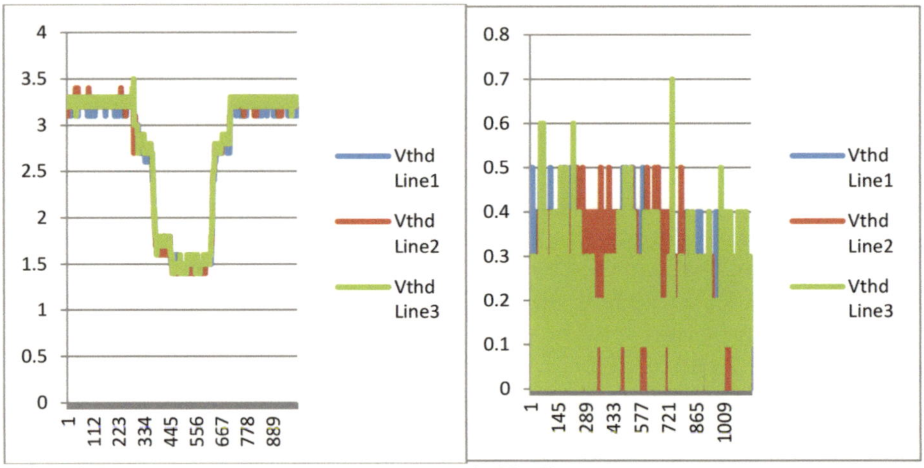

a. G1 - Master b. G2 - Slave

Figure 4.
The monitored voltage distortion factor on the three supply phases (THD-U).

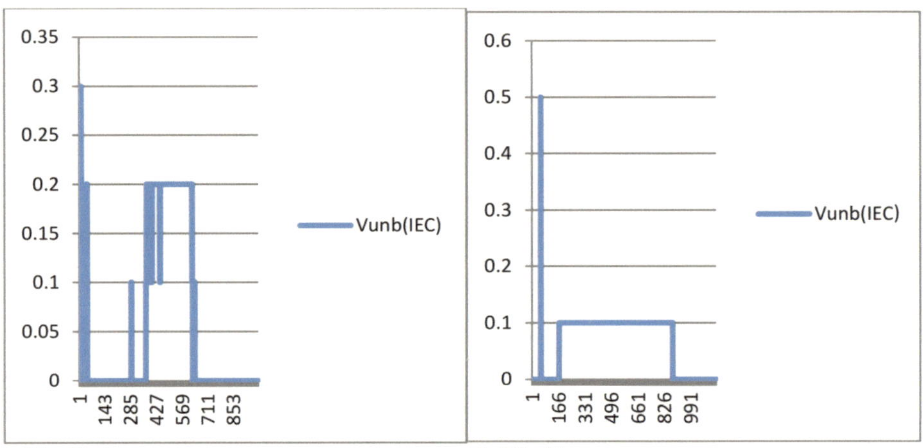

a. G1 - Master b. G2 - Slave

Figure 5.
The negative unbalance factor of the voltage at the TGD feeder bars (ks).

where I^- is the negative (reverse) sequence component of electrical currents and I^+ is the positive (forward) sequence component.

The automatic recloser is a very advanced outdoor (usually) three-phase circuit breaker with automation and electronic remote control, generally being a vacuum-breaking circuit breaker, possibly SF6 gas or air insulated, which is located in a stainless-steel tank, sealed for life, thus ensuring a long life with low maintenance [4, 5]. It features a fully insulated wiring system, full-wave-sized surge arresters with fixing supports and a "flexible" location, all of which contribute to a quick installation and a low specific cost. Actuation is done with coils whose operation is not necessarily based on the presence of the low-voltage (LV) supply.

The electronic automaton contains a microprocessor that can be programmed or programmed remotely. The program can be configured to have and offer different possibilities such as targeted protection and a variety of protocols such as RS, RTU, and

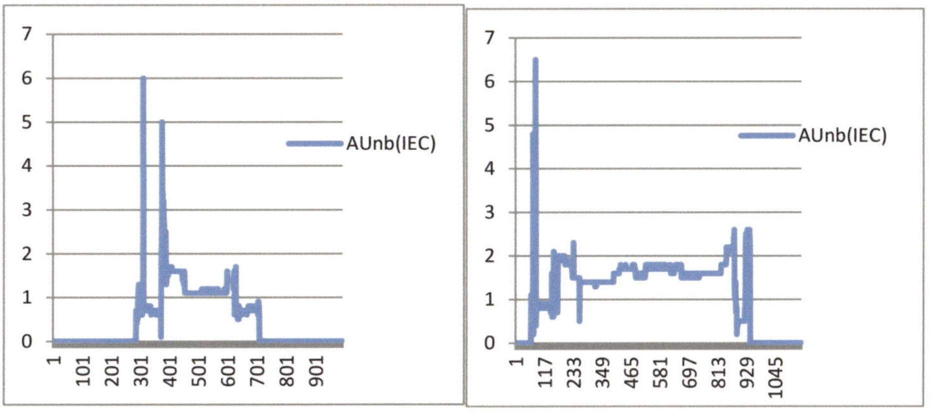

a. G1 - Master

b. G2 - Slave

Figure 6.
The variation of the unbalance factor of electric current of TGD (ks).

AMI. Thus, the recloser presents a specific property, which is called its "programming capability and bidirectional response" ("Software Capability"). The version of the program and its configuration determines the functionality of the automatic Recloser type.

The control electronics are housed in a stainless-steel control box designed for harsh climatic conditions. For the local operator, a control panel is provided for any type of climate, easily accessible by the "loco" or "remote" user. Remote control and monitoring can be provided without adding an RTU-type remote terminal unit. The Recloser circuit breaker uses interconnected vacuum breakers placed inside a stainless-steel tank filled with SF6 gas or air.

The use of this Recloser type equipment requires certain technical details such as:

- Type, typology, mode of operation, and other technical characteristics specific to the model.
- Windows Switchgear Operating System (the operating system of the switch under Windows).
- (WSOS) – Used to configure the recloser from a PC.
- Test and Training Set (TTS) – used to test control boxes.
- Specific Telemetry Protocol Implementations – For interfacing with remote control systems.
- Workshop & Field Test Procedures – Detailed and dedicated instructions on how to test the circuit breaker.
- Service Procedures (Service Procedures) – A set of instructions on how to remove and replace the electronic part of the machine in any situation (specific algorithmic presentation based on intervention scenarios).
- Declaration of conformity – the conformity requirements of the Standardization Institute of the region/state where the electrical equipment is located/mounted, previously sold (SR EN, CEI, STAS, NTE, etc.).

- Full technical specifications.
- Ambient conditions, specific to the installation/operational area, such as:
    - o Temperature during operation of −30 to +50°C.
    - o Humidity during operation is 0 to 100%.
    - o Solar radiation during operation max. 1.1 kW/m^2.
    - o Operating altitude of maximum 3000 m.
    - o The temperature range depends on the version of the control box.
    - o In the case of altitudes above 1000 m, the parameters must be declared according to ANSI-C37.60.

For Group G2, the supply bar voltages exhibit a waveform close to a sinusoid, characterized by a relatively low THD-U distortion factor (**Figure 4b**), which is within the permissible limits for low-voltage supply bars. The THD-U values range between 2.3 and 2.4% (**Figure 4b**). **Figure 5b** shows the unbalanced factor during the recording period. It is observed that the values of the negative unbalanced factor do not exceed the range of admissible values (<2%) at TGD. The analysis of the data in **Figure 7b** shows us that in 90% of cases, Pst admitted = 0.9 and Plt admitted = 1.0. The variations of these electric currents are presented in **Figure 7b**. The variations of the total electric current distortion factor – THD-I can be found in **Figure 8b**. The variation of the electric current unbalance factor, during the recording period, is shown in **Figure 6b** [1, 5, 8].

It is noticeable that the voltage level on phases and between phases is well within the normalized parameters (within the normalized/standardized limits) [8].

Figure 5 illustrates the variation of the unbalanced factor over the recording period [5, 8].

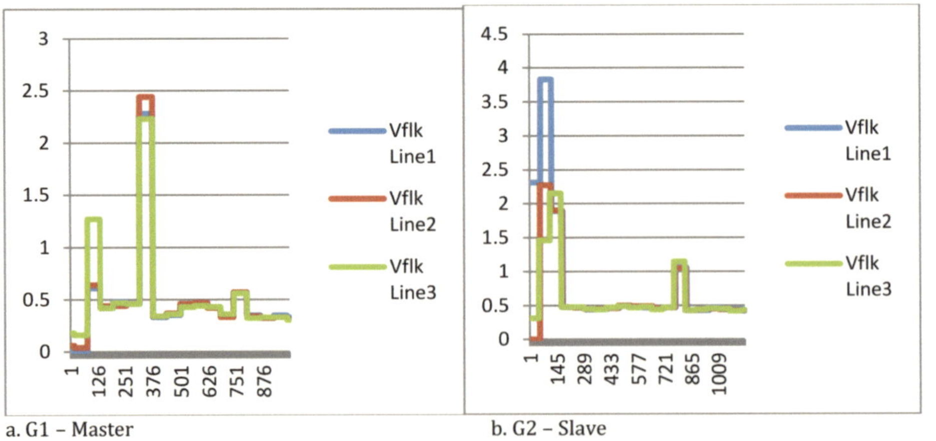

a. G1 - Master

b. G2 - Slave

Figure 7.
The variation in the level of voltage fluctuations at the TGD power bars.

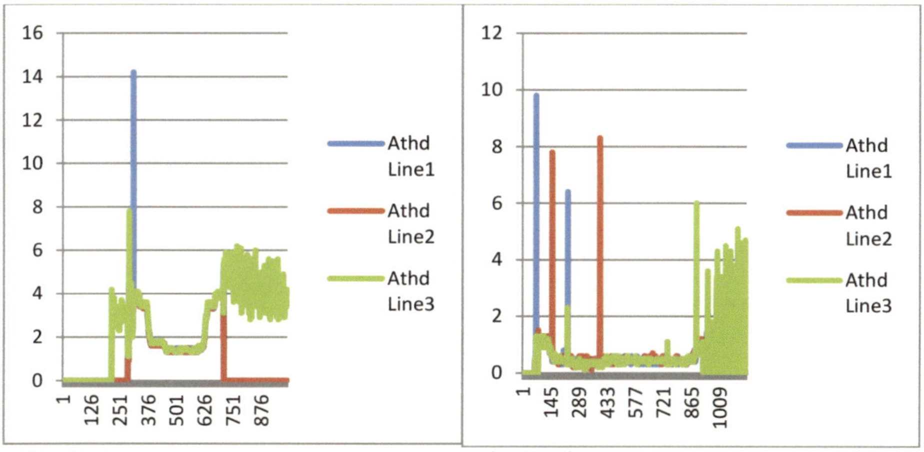

a. G1 – Master b. G2 – Slave

Figure 8.
The electric current distortion factor on the supply phases a, b, c (THD I) - monitored values.

Voltage fluctuations lead to the flicker effect, characterized by the indicators Pst (short-term factor) and Plt (long term factor).

$$P_{lt} = \sqrt[3]{\frac{1}{12} \cdot \sum_{j=1}^{12} P_{st}^{3} j} \tag{2}$$

The analysis of the data in **Figure 7** reveals that at the low-voltage bars of the receiver power system, the permitted levels of voltage fluctuations in the public network (P_{st} = 0.9 and P_{lt} = 1.0) are often exceeded. The number of events exceeding these limits is minimal and attributable to specific processes. Furthermore, in other users' energy systems, the permissible values differ from those in the public network, determined by the impact on their processes. The data in **Figure 7** show that phase B experiences larger voltage variations and higher flicker indicator levels. It is noted that the P_{st} and P_{lt} curves from the general connection table (TGD) within the company premises exhibit nearly identical shapes. Additionally, the measurements indicate that the flicker levels (P_{st}, P_{lt}) at the TGD power bars remain within normalized limits.

The electric currents on the three phases of the TGD have fairly close values, indicating, however, an asymmetric charging of the phases. The variations of these currents are shown in **Figure 9** [4, 9, 10].

From the analysis of the electrical measurements carried out, graphically represented previously, a series of observations emerge, such as the fact that the flicker level varies quite strongly and does not fall within the optimal range of values according to the CEI61000-3-4 standard (maximum 0.80 and 1.00 for specific long-term and short-term indicators in the case of low-voltage electrical networks) [4]. Under these conditions, the goal is to enhance power factor compensation while simultaneously improving flicker levels and optimizing the harmonic performance of both current and voltage.

From the analysis of the electrical measurements carried out, graphically represented previously, a series of observations emerge, such as those that there is an imbalance in the charging of the three phases of the electrical circuit analyzed, and it is interesting to observe, in this case, the balanced charging of the three monitored phases [4, 7, 11].

Figure 9.
The curve of electric currents on the power bar for TGD.

Thus, the analysis of the operation of the Diesel 2 generator group (G2) is done in parallel with that of the G1 group in the Master-Slave system; however, we must also know that there are a number of newer and older generating sets in number of six (G3, G4, G5,…., G8), existing on the analyzed user's platform, covering a good part of the load curve until close to its peak power of 7.5 MW which did not come into operation during the study period, and which, from a technical point of view, show a similar behavior, located in the geometric centers of consumption, in various areas on the entire analyzed industrial platform, being generally grouped by two or three, but the remaining six, are not the subject of this work (this chapter) [5].

The total electric current distortion factor—THD I (**Figure 8**)—indicates a sinusoidal current waveform in the supply circuit, as evidenced by the shape of the curves on the graphs. A preliminary comparison with voltage distortion data on the bars shows no direct correlation, particularly in phase a, where the harmonic distortion of the current is higher than in the other two phases.

It is important to note that the abrupt changes in the harmonic spectrum are caused by sudden variations in current, such as those occurring when motors and other electrical equipment are started (inrush current). These fluctuations are not indicative of the harmonic distortion of the electric current [7, 11, 12].

Figure 6 demonstrates that, during the recording period, the electric current unbalance factor remained within normal operational values, which are primarily influenced by the voltage unbalance in the power bars.

A comparable real-world scenario was implemented domestically at a large industrial facility. When a 110 kV supply cable caught fire, it resulted in an approximately 30-day outage for repairs. During this interruption, electric generators were installed in specific low-voltage sections. These generators operated for durations ranging from a few seconds or hours to several days within an automated "Master-Slave" system, which included a Recloser serving as an Automatic Reserve Activation (ARA) mechanism.

The electric currents on the three phases of the TGD are relatively similar in value but still show some asymmetry in phase loading. The variations in these electric currents are illustrated in **Figure 6** [1, 3, 7].

To determine the maximum active power for an industrial user, several methods are commonly employed:

- Demand coefficient method
- Binomial formula method
- Statistical method
- Graded load curve method

Figures 10 and **11** depict the variation in active power (P) and reactive power (Q) on the low-voltage (LV) bars, respectively. **Figure 12** shows the variation in apparent power transmitted through the circuit, while **Figure 13** provides a comparative view of the three-phase power variations [1, 2, 7].

The phase shift between the monitored voltages and electric currents at the power supply bar of the analyzed electrical equipment is shown in **Figure 14**.

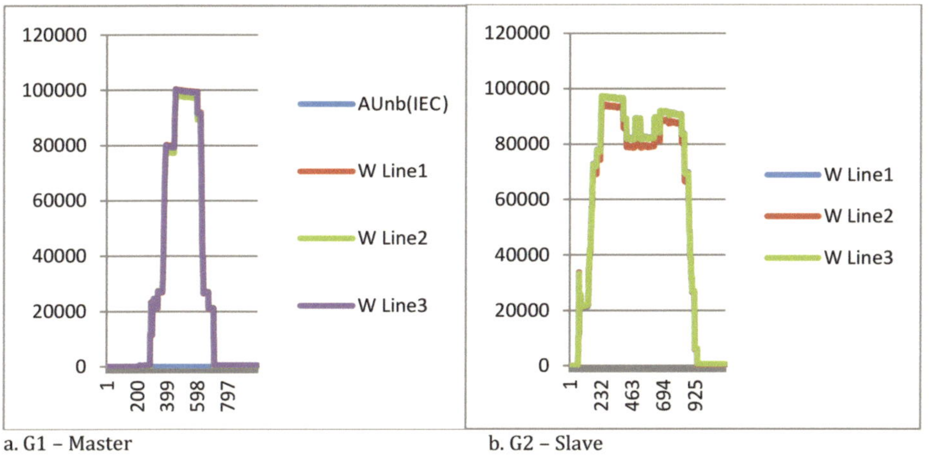

a. G1 – Master b. G2 – Slave

Figure 10.
The variation of P [W] on the LV bars of TGD.

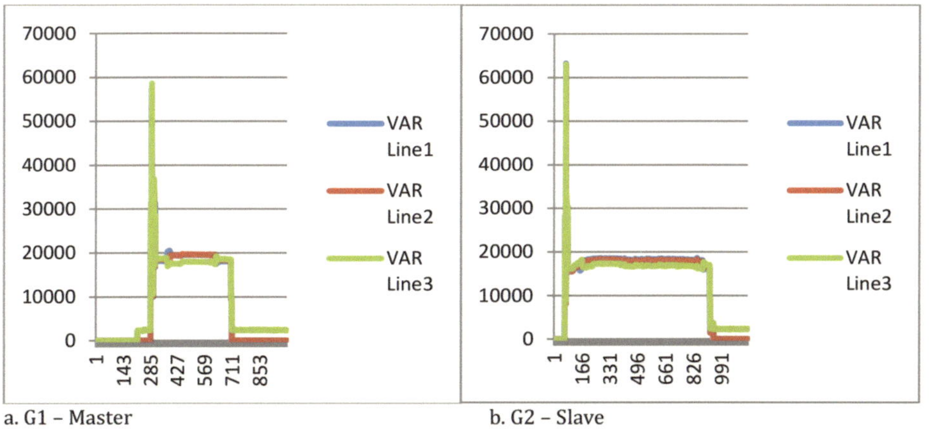

a. G1 – Master b. G2 – Slave

Figure 11.
Variation of Q [VAr] on the LV bars of TGD.

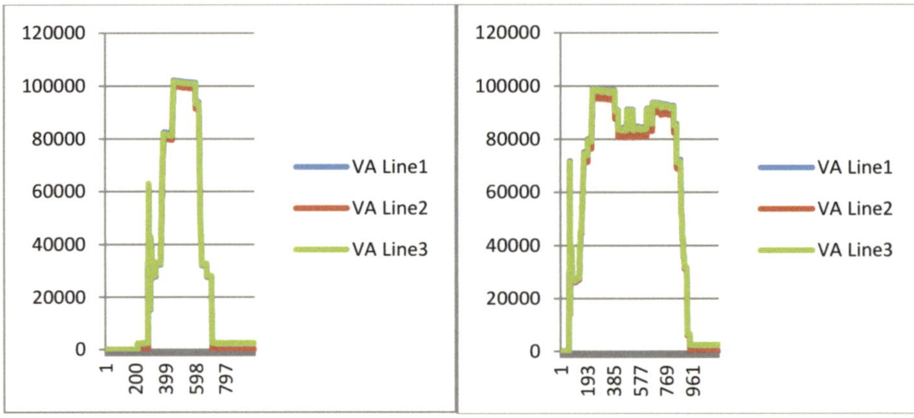

a. G1 - Master
b. G2 - Slave

Figure 12.
Single-phase apparent powers transited through TGD – S [VA].

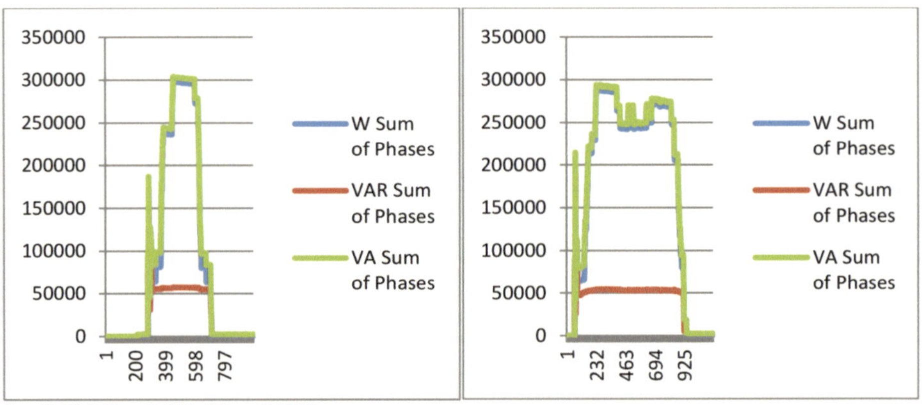

SumP [W], SumQ [VAr], SumS [VA]
a. G1 - Master
b. G2 - Slave

Figure 13.
Three-phase powers transited through TGD.

For Group G2, **Figure 10b** and **11b** illustrate the variation of active power (P) and reactive power (Q) on the LV bars, respectively. **Figure 12b** shows the variation in apparent power transmitted through the circuit, and **Figure 13b** presents a comparative view of the three-phase power variations. The phase shift between the monitored voltages and electric currents at the power supply bar for this group is detailed in **Figure 14b**.

The analyzed user presents a series of sub-consumers (grouped into three standard, priority and critical categories), which, in practice, are thus supplied in three ways or on three different electrical networks according to the Technical Connection Notice (ATR). It has a maximum contracted power of 7.5 MW (at peak) and presents a series of sub-consumers that have various CAEN codes of activities, i.e. they have technical-economic concerns such as [1, 7, 11, 12]:

- transportation;

DOI: http://dx.doi.org/10.5772/intechopen.1007189

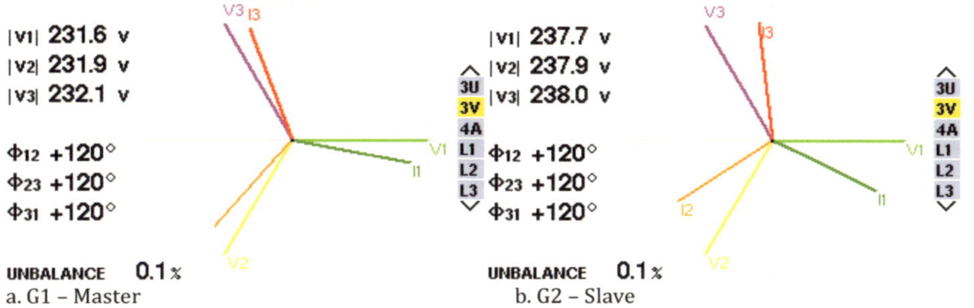

a. G1 – Master b. G2 – Slave

Figure 14.
The phase shift between voltages and electric currents at the supply bus on TGD.

- commercialization of horeca food products;
- restaurants and Dining Halls;
- conference rooms;
- accommodation;
- buildings with IT-software premises;
- buildings with office premises;
- research/design workshops;
- buildings with premises intended for FIREFIGHTERS or other similar private or state bodies/entities;
- cleaning company coordination services;
- cleaning services;
- production and sale of non-food products made of wood and with/from metal structures;
- production and marketing of products with complex structures;
- large cold storages with/without special equipment (on all types of national and international classification according to temperature, security, fragility, modes/modes of transport, etc.);
- marketing of non-food products including fuels or other volatile products;
- Hall services;
- repair and paint shops for large and very large machines;
- greenhouses;
- Hydrophor stations with drilled wells;

- thermal and electrical plants;
- and so on.

The average active power is determined by different calculation methods, for different periods of the year and for a whole year of the most demanded period of the day:

1. The coefficient of use method;
2. Specific energy method;
3. The method of the annual coefficient of use;
4. Maximum coefficient method.

The power circulated on a transmission line is directly influenced by the phase angle difference and inversely proportional to the reactance value. Series capacitors have the effect of reducing the total reactance of the transmission line [7, 11, 12]:

$$P = \frac{|V_1||V_2|}{X_l - X_c}\sin(\delta_1 - \delta_2) \quad (3)$$

For Group G2, the power factor variation over time is shown in **Figure 15b**. In **Figure 16b**, you can see the values of the voltage harmonics on the three phases (A, B, and C) at the power bars for TGD, and in **Figure 16b**, you can see the values of the current harmonics on the three phases (A, B, and C) at the power bars for TGD [7, 11, 12].

Figure 15 shows the power factor variation for each phase during equipment operation, indicating significant reactive power circulation in the supply circuit.

From the analysis of the electrical measurements carried out, graphically represented previously, a series of observations emerge, such as the fact that the level of the power factor varies permanently, presenting its values that are not always found in the range of 0.90 and 1.00, i.e., between the neutral value and one, this leads to penalties

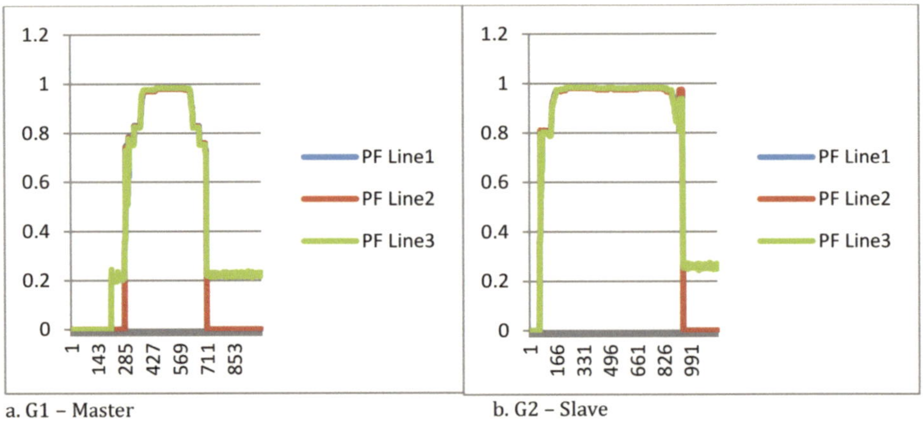

Figure 15.
Variation of the power factor at the power bars for TGD on phases A, B, C.

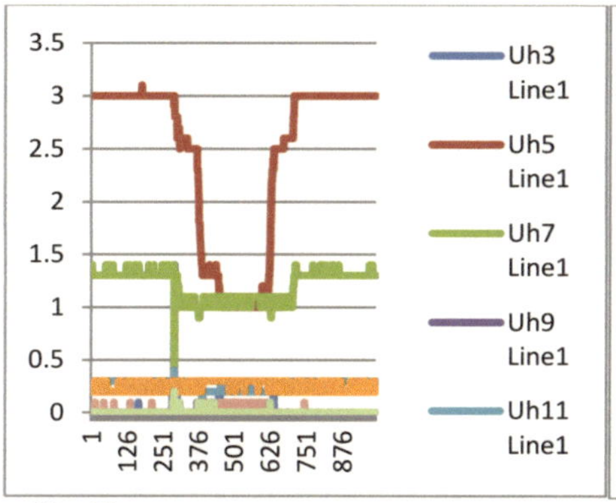

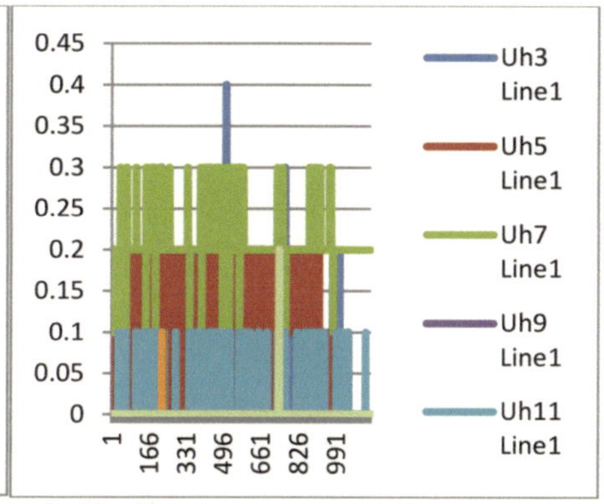

a. G1 – Master b. G2 – Slave
A. Harmonic level [%] phase 1 (voltages)

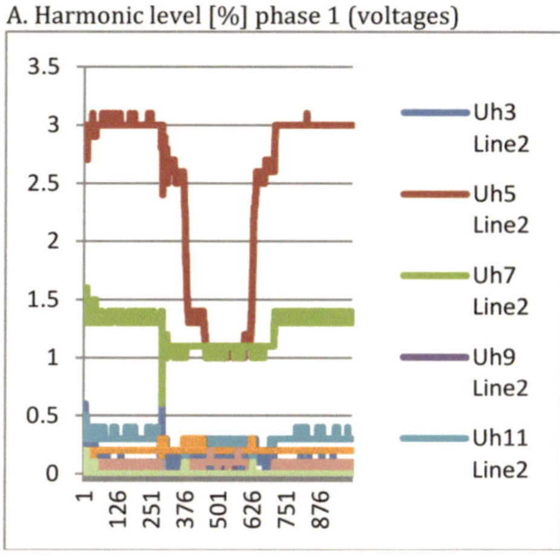

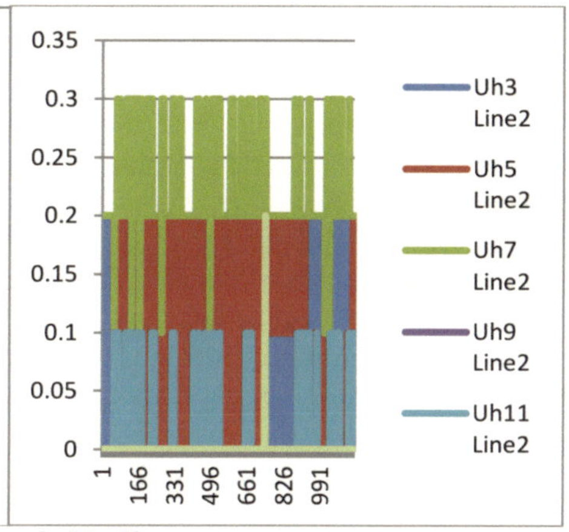

a. G1 – Master b. G2 – Slave
B. Harmonic level [%] phase 2 (voltages)

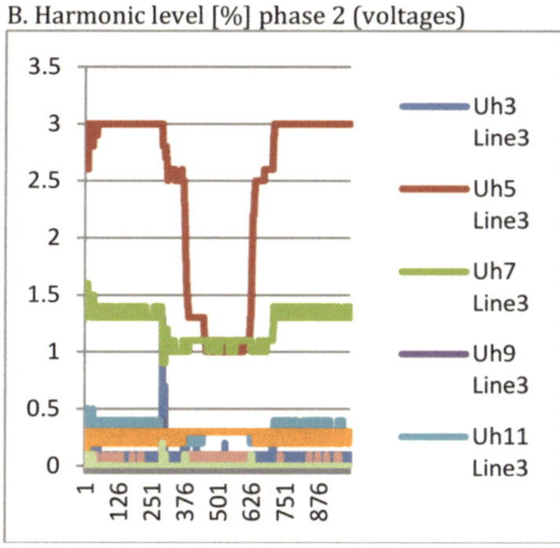

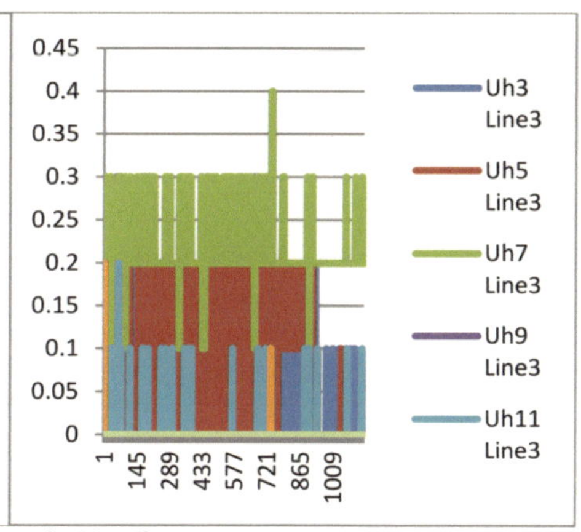

a. G1 – Master b. G2 – Slave
C. Harmonic level [%] phase 3 (voltages)

Figure 16.
The monitored level of voltage harmonics on the three phases for TGD.

in the supplier+distributor+user relationship [5, 8, 12]. Under these conditions, the aim is to improve the power factor compensation, bringing it between the values of 0.90 and 1.00.

The recorded harmonic spectrum values are influenced by the operating conditions, including the frequent starting and stopping of motors and other electrical equipment such as lighting. To accurately assess the harmonic spectrum, only the values obtained during periods of steady operation should be considered.

In **Figure 16**, we can see the values of the voltage harmonics on the three phases (A, B, and C) at the supply bars for TGD. In **Figure 17**, we can see the values of the current harmonics on the three phases (A, B, and C) at the supply bars for TGD [5, 8, 12].

Nonlinear phenomena, especially certain types of bifurcations, have been identified as responsible for a number of stability problems in isolated power microsystems. In particular, the lack of post-failure equilibrium points, commonly associated with singular bifurcation points (SNBs) and certain types of limits produced bifurcations (LIBs), has been identified as the main cause of voltage collapse. There are explicit examples in the literature of these bifurcations in power systems and their connection to voltage stability, so essentially bifurcation points are balance points where the "quantity" and/or "quality of balance" changes, in relation to a nonlinear system of dynamic equations, during the slow variation of the system parameters [5, 8, 12].

The analysis of the electrical measurements, as graphically presented, reveals several key observations. Notably, the level of electric current harmonics is relatively high, with harmonics at orders 3, 5, 7, and 9 exhibiting particularly elevated values. In light of these findings, the objectives are to enhance power factor compensation, mitigate flicker levels, and optimize the harmonic performance of the electrical system.

The performance standard regulates the commercial quality criteria defined by the performance indicators for ensuring the electricity supply service to consumers (users), as well as the reporting requirements for suppliers [5, 8, 12].

The standard applies in the relations between distribution operators and those RED users who simultaneously meet the conditions:

a. have the installations at the nominal alternative voltage in the range 0.4 – 110 kV and at the nominal frequency of 50 Hz;

b. through their operating regime, they do not introduce disturbances in the supply of other RED users in the area;

c. falls within the maximum power approved by the technical approval for connection and complies with the conditions stipulated in the distribution/supply contract.

Art. 22. (1) The voltage distortion factor at LV and LV must be less than or equal to 8%. (2) Under normal operating conditions, the harmonic voltages at the demarcation points, at LV and MV, must not exceed the maximum limits indicated in the table below, for 95% of the week.

For a stricter interpretation of the results, it is specified that for making these comparisons, Law no. 13/2007, amended by Law 123/2012, Performance Standard for electricity distribution services RO_ANRE_Code 28.1.013.0.0.30, was used.08.2017 modified by Ord._ RO_ANRE no. 11/2016 and Internal Technical Specifications of distribution operators (or Technical Sheets – **Table 1**).

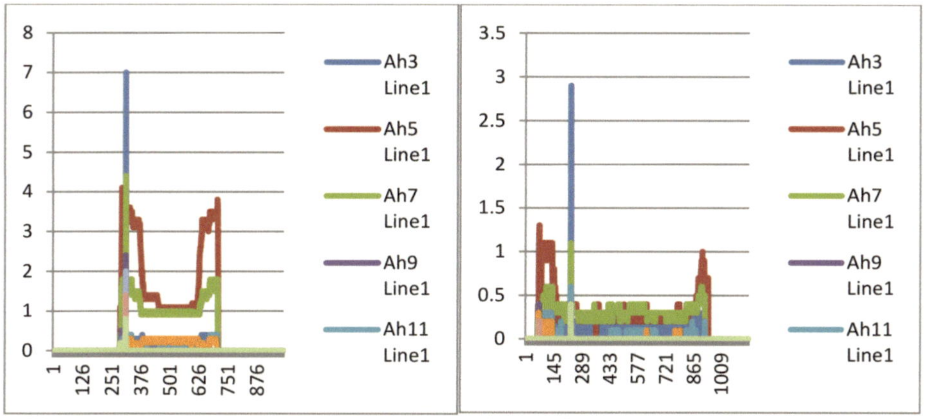

a. G1 – Master b. G2 – Slave
A. Harmonic level [%] phase 1 (electric currents)

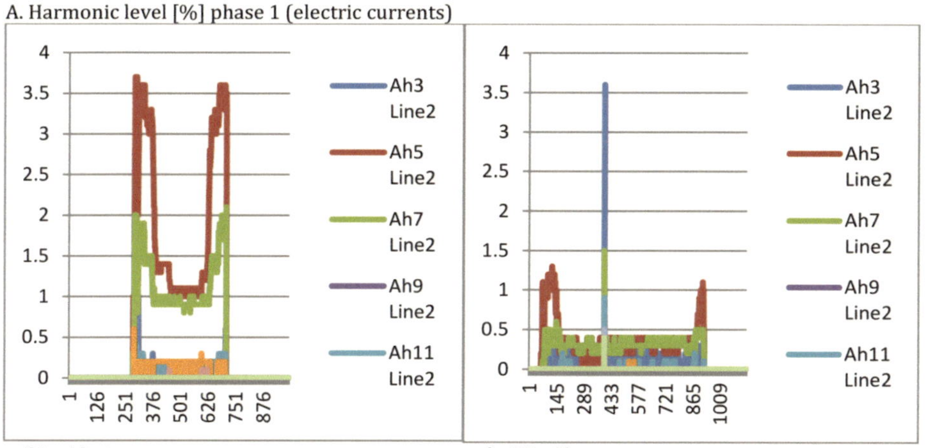

a. G1 – Master b. G2 – Slave
B. Harmonics level [%] phase 2 (electric currents)

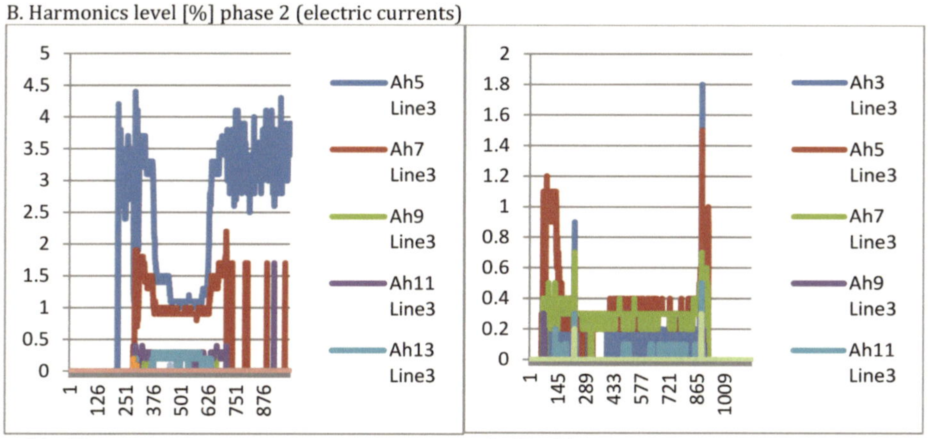

a. G1 – Master b. G2 – Slave
C. Harmonics level [%] phase 3 (electric currents)

Figure 17.
The level of electric current harmonics on phases A, B and C at the TGD power bars.

The inclusion in the performance standard of some energy quality elements of the G1 group (**Table 2**) [5, 8, 12] - We must also take into account a series of minimal descriptions, considered useful in support of the proposed ones.

Odd harmonics (% of fundamental):				Even harmonics (% of fundamental):	
Not a multiple of 3		Multiple of 3			
Rank	Threshold	Rank	Threshold	Rank	Threshold
5	6%	3	5%	2	2%
7	5%	9	1.5%	4	1%
11	3.5%	15 and 21	0.5%	6 to 24	0.5%
13	3%				
17	2%				
19, 23, 25	1.5%				

Table 1.
Internal technical specifications of distribution operators (or Technical Data Sheets).

Odd harmonics (% of fundamental):			
Not a multiple of 3		Multiple of 3	
Rank	Threshold [%]	Rank	Threshold [%]
5	5.4%	3	6.9%
7	4.7%	9	4.1%
11	3.7%	15	3.4%
13	3.6%	21	1.2%
17	2.5%		
19	1.3%		
23	0.4%		
25	0.1%		

Table 2.
The inclusion in the performance standard of some energy quality elements of the G1 group.

Thus, this analysis is carried out on a real case, and becomes more and more useful as, over the course of a year, more and more such requests appear, sometimes unpleasant incidents of the type mentioned above, in which the sub-consumers this complex user must cope with without affecting their specific activities.

According to the Performance Standard that regulates the commercial quality criteria defined by the performance indicators for ensuring the electricity supply service to consumers/users, as well as the reporting requirements for suppliers, making a comparison between the limits imposed by the Performance Standard and the values of the monitored harmonics on the electrical network connection circuit of the electric generator of the diesel group G1 under analysis (on the LV side), it is observed that the latter does not fall within the upper maximum admissible limits in the normative values. See rank harmonics:3,9,11,13,15,17,21 highlighted in red.

The inclusion in the performance standard of some energy quality elements of the G2 group (**Table 3**) [5, 8, 12] - According to the Performance Standard that regulates the commercial quality criteria defined by the performance indicators for ensuring the service of supplying electricity to consumers/users, as well as the requirements

Odd harmonics (% of fundamental):			
Not a multiple of 3		**Multiple of 3**	
Rank	**Threshold [%]**	**Rank**	**Threshold [%]**
5	5.9%	3	6.2%
7	4.4%	9	4.4%
11	3.7%	15	3.1%
13	2.6%	21	1.9%
17	1.5%		
19	1.3%		
23	0.7%		
25	0.4%		

Making a comparison between the limits imposed by the Performance Standard and the harmonic values for the electric generator of the G2 diesel group under analysis (on the LV side), it can be seen that the latter does not fall within the upper maximum admissible limit in the normative values. See rank harmonics: 3,9,11,15,21.

Table 3.
The inclusion in the performance standard of some energy quality elements of the G2 group.

of reporting for suppliers, making a comparison between the limits imposed by the Performance Standard and the harmonics values monitored on the electrical network connection circuit of the electric generator of the G2 diesel group under analysis (on the LV side), it is observed that the latter do not falls within the upper maximum admissible limits in the normed values. See rank harmonics:3,9,11,15,21 highlighted in red.

Because the Master-Slave automation is done with the help of a Recloser, we will say a few words about this equipment as well [8].

3. Elements regarding the photovoltaic charging system of G1 and G2 groups accumulators

The proposed photovoltaic system is sized to provide the necessary electrical energy for internal services, protections, reclosers, programmable automata, but also for the maintenance of consumers for periods of a few seconds until the diesel generating sets start [8]. Of course, this is possible by feeding or rather, by connecting this photovoltaic system to the accumulators of the diesel groups. Another function of the accumulators of these groups is to keep them warm (keeping the engine blocks and the oil warm to be able to make a quick and correct start whenever necessary, especially in the Slave group) between 42 and 50°C.

The operation of the photovoltaic system is presented below on a typical day. It is anticipated that it will work similarly in 75% of cases. In the short, dark days, it is possible to even use the groups to charge the batteries or, of course, the electrical network if we can work and connect and not in the island.

The voltage at the power supply bars of the receivers shows variations (222.9... 241.8) Voltage per phase/ (383.8...417.6) Voltage between phases, for TGD (**Figure 18**), which must fall within the limits of 230 ± 10% /400 ± 10% V (207...253/360.../440 V). Rapid fluctuations in voltage values, caused by specific processes within the enterprise, result in recorded voltage variations accompanied by the flicker effect.

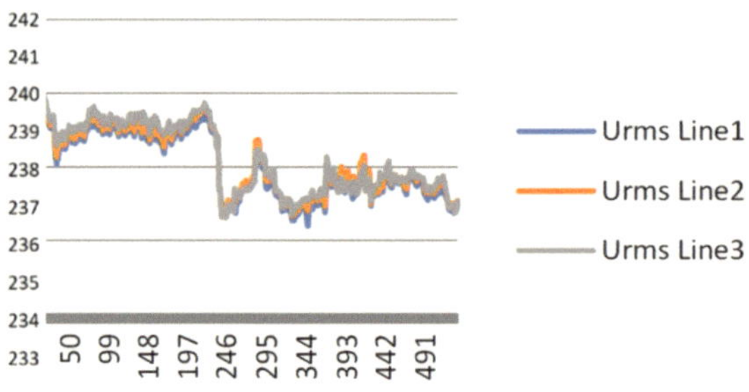

Figure 18.
Busbar voltage curve for TGD-PV between phases.

The three phases are loaded almost equally. It is observed that the supply voltage values during the measurement period remained within the normative limits [3, 5–7].

The voltages from the supply bars exhibit a waveform close to a sinusoid, characterized by a relatively low total harmonic distortion voltage (THD-U) factor (**Figure 19**). This distortion factor is within the permissible limits for low-voltage supply bars (THD-U_{admis} of 8%). Specifically, the THD-U values are relatively low, ranging from 1.1 to 3.4% (**Figure 19**) [3, 6, 7].

Figure 20 shows the evolution of the unbalanced factor throughout the monitored interval [5, 8].

The analysis of the data presented in **Figure 21** reveals that, at the low-voltage bars of the receiver power system, the permissible levels of voltage fluctuations are often exceeded relative to the public grid standards (P_{st} = 0.9 and P_{lt} = 1.0). Although

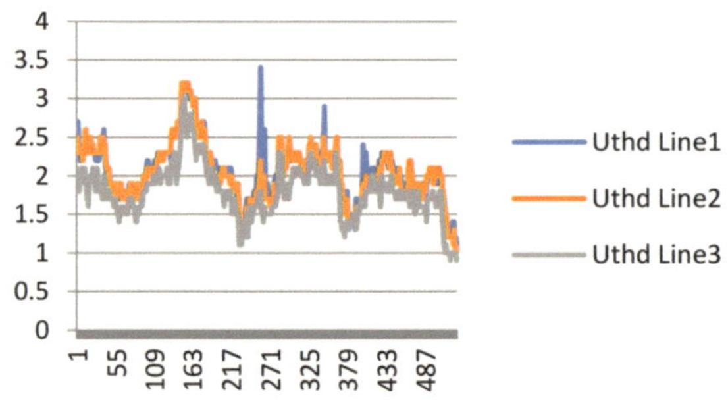

Figure 19.
The monitored voltage distortion factor on the three supply phases (THD-U).

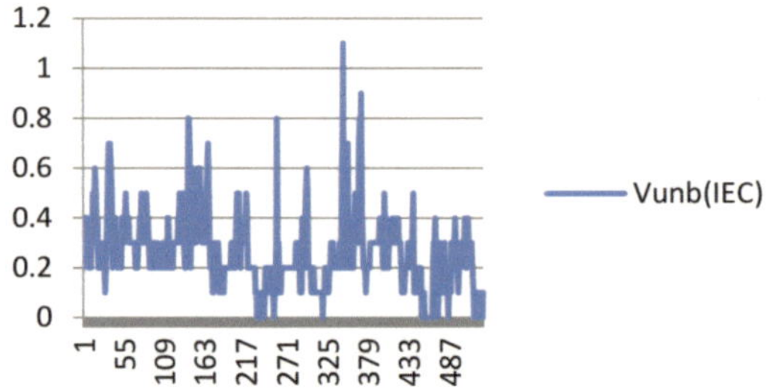

Figure 20.
Unbalance factor of the voltage at the TGD-PV feeder bars (ks).

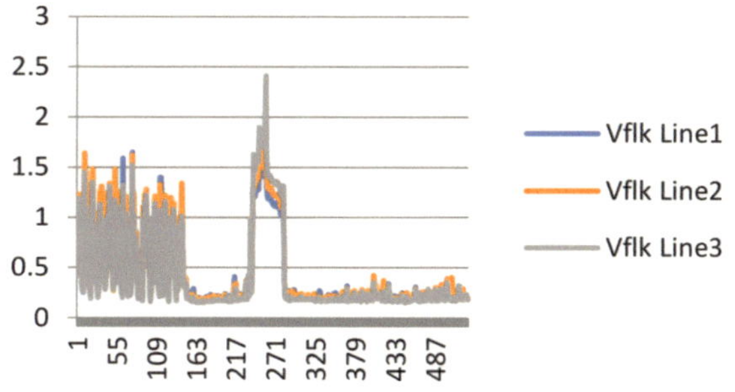

Figure 21.
The variation in the level of voltage fluctuations at the TGD-PV power bars.

instances where values exceed these limits are infrequent, they are attributed to specific processes. Additionally, the permissible values in other users' energy systems differ from those in the public network, as they are tailored to the specific requirements of their processes. It is noted that the P_{st} and P_{lt} curves, from the general connection table (TGD) within the company's premises, followed, have very similar shapes (alliance).

Also, based on the measurements, it is observed that at the TGD power bars, the flicker level is within the normalized parameters (limit). Causes of voltage fluctuations in photovoltaic systems:

- Variability of solar radiation: Rapid changes in weather conditions (e.g. passing clouds) can cause rapid variations in the power generated by PV panels.

- Network Connections: Sudden addition or withdrawal of PV power to the grid can lead to voltage fluctuations.

- Variable tasks: The variable energy consumption of consumers can also contribute to voltage fluctuations in grids with PV integration [13].

The electric currents across the three phases of the TGD are relatively similar, yet they indicate some degree of phase asymmetry. These current variations are illustrated in **Figure 22** [5, 8]. It is worth noting once again that the photovoltaic system accommodates all eight groups.

The total electric current distortion factor—THD-I (**Figure 23**)—suggests a sinusoidal current waveform in the supply circuit. Initial comparisons with voltage distortion data show no direct correlation, particularly in the second phase, where

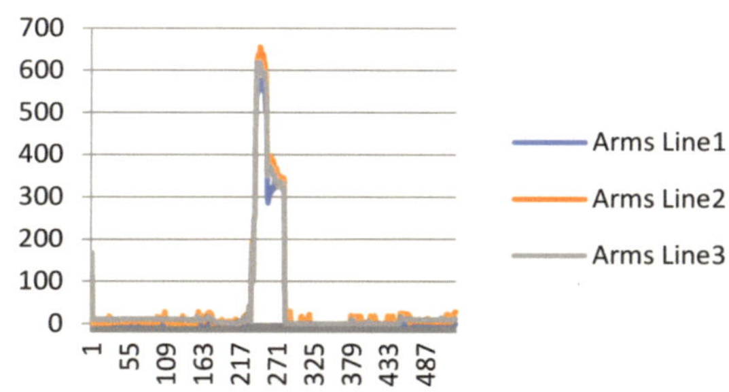

Figure 22.
The curve of electric currents on the power bar for TGD-PV.

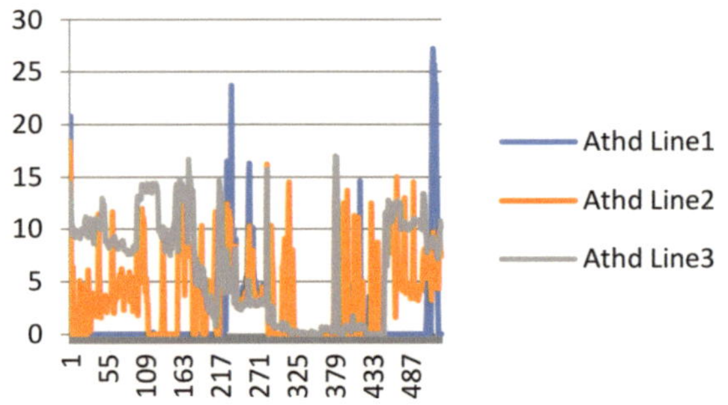

Figure 23.
The monitored electric current distortion factor on the three supply phases (THD I).

current harmonic distortion is higher than in the other phases. It is important to note that abrupt changes in the harmonic spectrum are caused by sudden current variations during equipment startups (inrush current), which are not indicative of harmonic distortion for analysis purposes.

Figure 24 illustrates the variation in the electric current unbalance factor over the recording period. **Figure 25** shows the variation in active power (P), while **Figure 26** depicts the variation in reactive power (Q) on the low-voltage (LV) bars for accumulators (TGD-PV). **Figure 27** presents the variation in apparent power transmitted through the circuit, and **Figure 28** offers a comparative view of the three-phase power variations [1, 3, 6, 12].

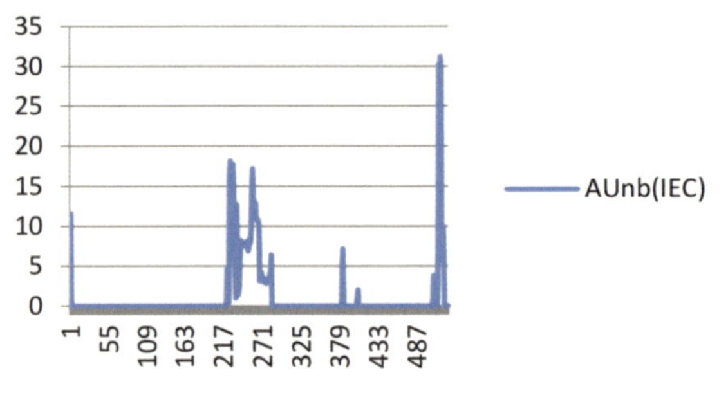

Figure 24.
The variation of the negative electric current unbalance factor of TGD (ks) - Aunb [%].

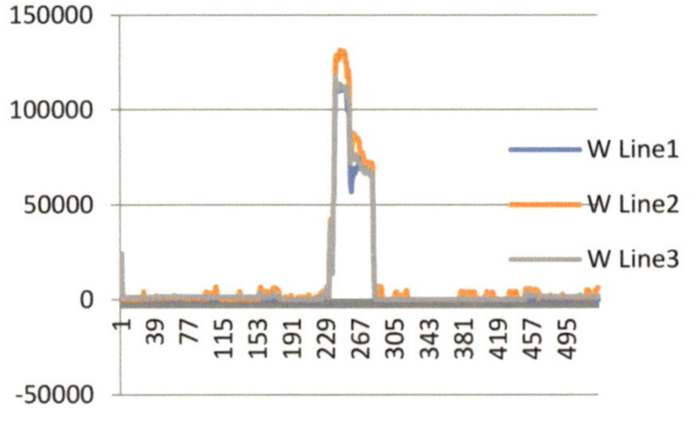

Figure 25.
Variation P (active powers).

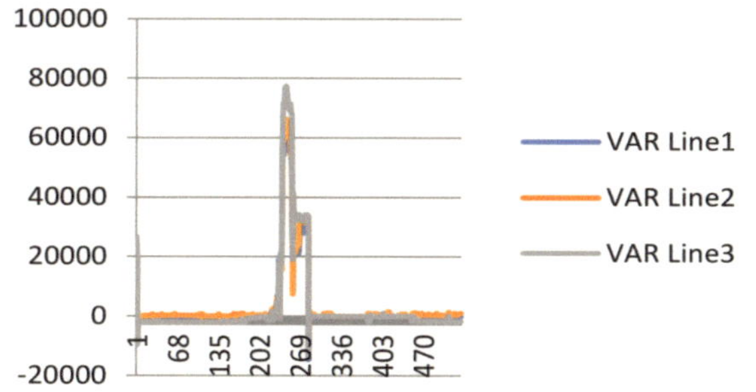

Figure 26.
The variation of Q [VAr] on the LV bars of TGD-PV.

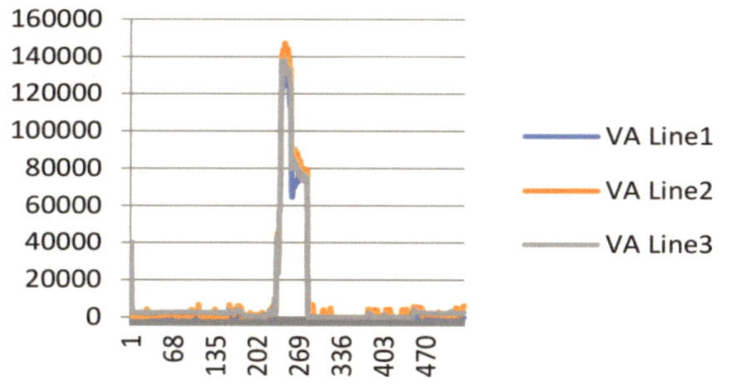

Figure 27.
Single-phase apparent powers transited through TGD-PV.

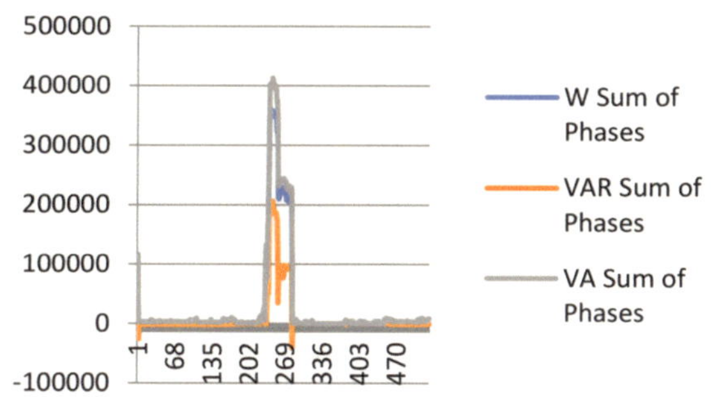

Figure 28.
Three-phase powers transited through TGD-PV (SumP [W], SumQ [VAr], SumS [VA]).

During the shadow period, the photovoltaic panels become consumers. As a rule, they consume between 1 and 5% of the installed power. By ensuring efficient management of voltage and reactive power, these devices contribute to more efficient integration of solar energy into electrical networks, thus supporting the transition to renewable energy sources and reducing environmental impact [1, 3, 7].

Figure 29 displays the instantaneous phase shift between the monitored voltages and electric currents at the power supply bar of the analyzed electrical equipment [3, 6, 12].

Figure 30 illustrates the variation in the power factor for each phase during the recording period and operational intervals of the equipment. The power factor is calculated using Eq. (4). The graph clearly indicates a significant presence of reactive power circulating within the supply circuit [3, 6, 7].

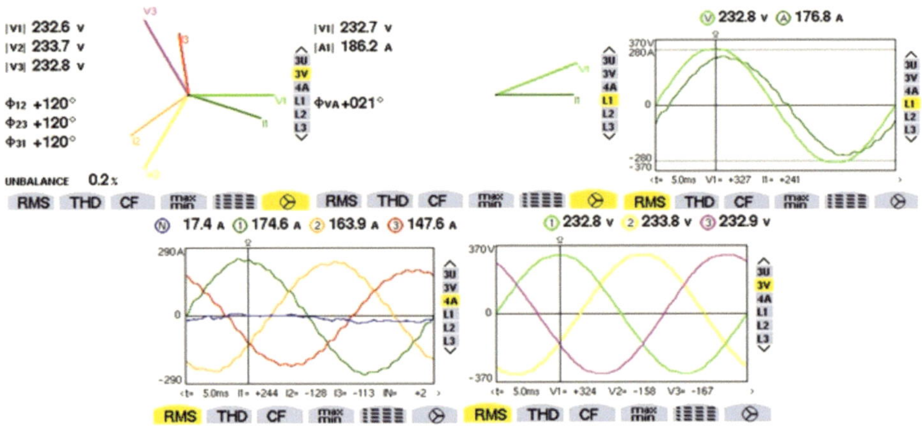

Figure 29.
The phase shift between voltages and electric currents at the power bar on TGD-PV – peek.

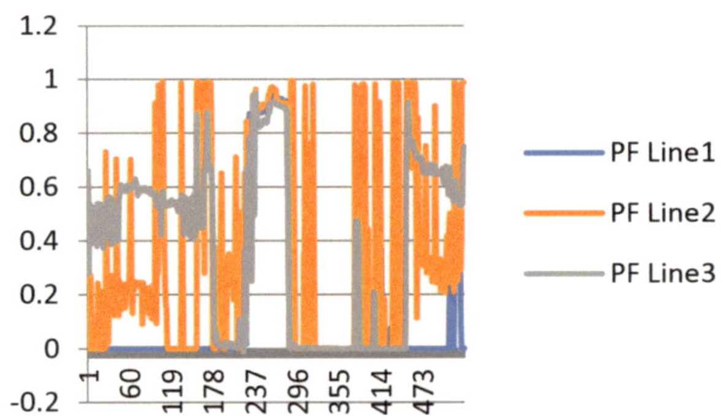

Figure 30.
The variation of the power factor at the power bars for TGD-PV on phases A, B, and C.

$$PF = \frac{DPF}{\sqrt{1+THDI^2}} = \frac{0,991}{\sqrt{1+0,15^2}} \cong 0,98 \qquad (4)$$

The recorded harmonic spectrum values are influenced by the specific operating conditions, including frequent startups and shutdowns of motors and other electrical equipment, such as lighting. To accurately assess the harmonic spectrum, only values obtained during steady-state periods should be considered.

In **Figure 31**, the values of the voltage harmonics on the three phases (A, B, and C) at the power bars for TGD-PV can be observed [1, 6, 7].

In **Figure 32**, we can see the values of the current harmonics on the three phases (A, B, and C) at the power bars for TGD-PV [3, 6, 7].

Less than 1% of the Earth's atmosphere is made up of water vapor, carbon dioxide, ozone, methane, nitrous oxide, and sulfur hexafluoride, gases known as greenhouse gases [5, 13]. The first five gases listed above occur naturally and produce a natural greenhouse effect, capable of keeping the global temperature 30°C higher than in their absence, thus supporting life. The concentration of greenhouse gases is increasing, as a direct result of human activity.

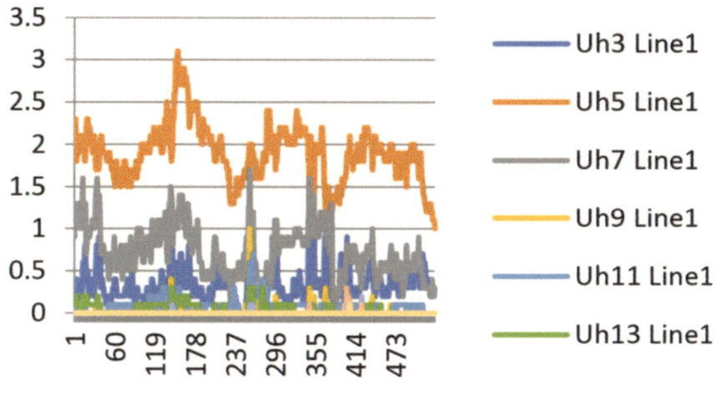

Figure 31.
The monitored level of voltage harmonics on the three phases for TGD-PV.

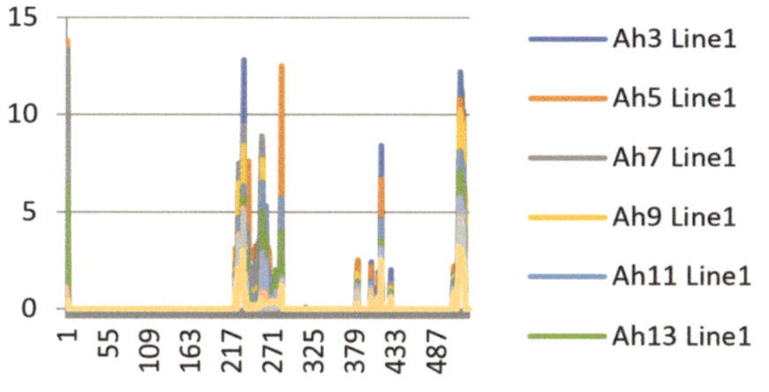

Figure 32.
The level of electric current harmonics on phases A, B, and C (similar) at the TGD-PV power bars.

No.	Measure	Expected energy		approximate cost of implementation
		MWh/ year	TOE/ year	[EURO]
1	Master-slave operation of G1 + G2 + PV + Accumulator + Recloser + Automation	97.5	8.385	40,000
Obs. TOE – Tone Oil Equivalent.				

Table 4.
System groups G1 and G2.

Elements	UM	Considered fuels		
		HFO	natural gas	coal
Specific emission of greenhouse gases	kg CO_2/MJ	0.0773	0.0561	0.0945
Reduction of energy losses	MWh/year		97.5	
Fuel consumption to produce 1 MWhe	MWh/year		273	
	MJ/year		982,800	
Reduction of greenhouse gas emissions	tons of CO_2/year	1.51941	8.27026	37.14984
Reduction of total greenhouse gas emissions	tons of CO_2/year		46.93951	

Table 5.
Analyzes of greenhouse gas emissions for System groups G1 and G2.

The year	Financial economy	The residual value	Income	Installation cost + equipment	Total expenses	Net cash flow	The internal rate of return represents the limit value of the discount rate for which the project does not produce losses	The discounted net income mathematically adds up all the discounted estimated payments and receipts	Capital recovery period with the difference that the amounts are no longer discounted
A	Eph	Vr	V	CM + E	Tc	FNN	IRR	VNA	t
1	5850.00		5850.00	40,000.00	40,000.00	−34,150.00		5318.18	6.68
2	6054.75		6054.75		0.00	−28,095.25		10,322.11	
3	6266.67		6266.67		0.00	−21,828.58		15,030.35	
4	6486.00		6486.00		0.00	−15,342.58		19,460.37	
5	6713.01		6713.01		0.00	−8629.57		23,628.62	
6	6947.96		6947.96		0.00	−1681.61		27,550.57	
7	7191.14		7191.14		0.00	5509.53		31,240.76	
8	7442.83		7442.83		0.00	12,952.37	−0.30	34,712.90	
9	7703.33		7703.33		0.00	20,655.70	−0.16	37,979.86	
10	7972.95		7972.95		0.00	28,628.65	−0.07	41,053.78	
11	8252.00		8252.00		0.00	36,880.65	−0.01	43,946.06	
12	8540.82		8540.82		0.00	45,421.48	0.04	46,667.43	
13	8839.75		8839.75		0.00	54,261.23	0.07	49,227.99	
14	9149.14		9149.14		0.00	63,410.37	0.10	51,637.24	
15	9469.36		9469.36		0.00	72,879.73	0.12	53,904.13	
16	9800.79		9800.79		0.00	82,680.52	0.14	56,037.07	
17	10,143.82		10,143.82		0.00	92,824.34	0.15	58,043.97	
18	10,498.85		10,498.85		0.00	103,323.19	0.16	59,932.28	

The year	Financial economy	The residual value	Income	Installation cost + equipment	Total expenses	Net cash flow	The internal rate of return represents the limit value of the discount rate for which the project does not produce losses	The discounted net income mathematically adds up all the discounted estimated payments and receipts	Capital recovery period with the difference that the amounts are no longer discounted
19	10,866.31		10,866.31		0.00	114,189.51	0.17	61,709.01	
20	11,246.63	1000.00	12,246.63		0.00	126,436.14	0.18	63,529.39	

Obs. If the amount of 97.5 MWh/year were to triple in one calendar year (that is, the groups would triple their operating times in 1 year), the investment recovery terms would decrease by about three times, at an energy price of around 200 euros/MWh [13].

Table 6.
Economic analysis for system groups G1 and G2.

The reduction of greenhouse gas emissions has gained, in the last decade, a privileged place in energy and environmental policies all over the world. The effects of climate change have become more and more visible, and combating them must become an absolute priority of all the countries of the world.

In order to determine the impact on greenhouse gas emissions of the main measures to increase energy efficiency, it is considered that the reduction of electricity losses is equivalent to the reduction of the same amount of electricity production in the SEN thermoelectric plants. They use as primary energy source the three categories of fossil fuels present in Romania: fuel oil, natural gas and coal. To estimate the specific greenhouse gas emissions of each type of fuel, the values from the energy prescriptions related to this field were considered.

In the tables below (**Table 4** [5]), you can find the analyses of greenhouse gas emissions (**Table 5** [8]) for a hypothetical case in which groups G1 and G2 produce 97.5 MWh/year (8.385 TOE/year), highlighting for each subsystem (electro) a series of environmental and economic indicators (**Table 6** [10]) with the related characteristic results [8, 12].

All the economic terms above can be found in specialized economic literature and, for reasons of copyright and anti-plagiarism, have not been presented in detail. This chapter is especially aimed at those who possess some technical and economic knowledge in this respect and have knowledge of cost-benefit analysis [11, 12].

4. Conclusions

The authors presented a case study on the efficient operation of smart electrical systems within the context of sustainable development, focusing on the isolated operation of a medium-voltage electrical network for specific users. The study included an analysis of the quality of the electricity supply in a three-phase circuit during the intervention of two diesel generators (G1 + G2) operating in a Master-Slave system alongside photovoltaic and accumulator systems. Results from actual measurements were interpreted and assessed against the required performance standards.

The study found that accidental or intentional isolation of the network, depending on needs and requests, can lead to significant voltage and frequency variations. Such variations can result in substandard electricity supply parameters, potentially causing considerable damage to users. Therefore, this mode of operation is generally undesirable and should be avoided. However, it is essential to adopt protective measures in real-time through auto-adaptive intelligent systems during inslanding to mitigate these effects.

The analysis also revealed that intentional isolation might be beneficial in certain scenarios, such as restarting operations for demanding industrial users after a major disturbance or during specific training maneuvers. Nonetheless, operating an isolated portion of the electrical network and serving critical sub-consumers presents complex challenges, particularly when dealing with variable loads and primary energy supplied by generator groups.

In summary, while isolated operation of medium voltage electrical networks can offer advantages under specific conditions, it requires careful management and real-time adaptive measures to ensure stability and minimize risks. Further research and practical applications are needed to optimize such systems for sustainable development.

Author details

Laurentiu Constantin Lipan[1*] and George Alexandru Florea[2]

1 National University of Science and Technology "POLITEHNICA" of Bucharest, Romania

2 Power and Lighting Tehnorob SA, Bucharest, Romania

*Address all correspondence to: laurentiu.lipan@upb.ro

References

[1] Alymov I, Averbukh M. Monitoring energy flows for efficient electricity control in low-voltage smart grids. Energies. 2024;**17**(9):2123. DOI: 10.3390/en17092123

[2] Marek W, Aneta C. Energy performance in residential buildings as a property market efficiency driver. Energies. 2024;**17**(10):1-18. DOI: 10.3390/en17102310. WOS:001232345100001, e-ISSN 1996-1073, IDS Number SC8G9

[3] Pîslaru-Dănescu L, Lipan LC, El-Leathey LA. Microgrids smart structures used for back-up power supply. EEA-Electrotechnics, Electronics, Automatics. 2016;**64**(3):68-74. Print ISSN:1582-5175

[4] Florea G, Lipan A, Lipan LC, Sebastian A. In: Dragan G, editor. Energ 08, Explanatory Dictionary of Overhead Power Lines. Bucharest: AGIR Publishing House/Romanian Academy Publishing House; 2007. 196 p. ISBN 973-8140-17-4/ ISBN 973-27-0747-X

[5] Eremia M, Liu CC, Edris AA. Advanced Solutions in Power Systems. Piscataway, NJ: IEEE Press, Wiley; 2016. p. 1042. ISBN 1119175399, 9781119175391

[6] Chousidis C, Nilavalan R, Lipan LC. Expanding the use of CTS-to-self mechanism for reliable broadcasting on IEEE 802.11 networks. In: Proceedings of the IEEE International Conference on Wireless Communications and Mobile Computing Conference (IWCMC), 4-8 Aug 2014. Nicosia, Cyprus: IEEE; 2014. p. 1051, 1056. DOI: 10.1109/IWCMC.2014.6906500

[7] Rezaei N, Pezhmani Y, Jordehi AR, Mansouri SA. A two-layer hybrid robust-stochastic model for energy management of isolated multi-energy microgrids with mobile storage systems and hydrogen refueling stations. Journal of Energy Storage. 2024;**90**(Part A):111905

[8] Ivan A, Moshe A. Monitoring energy flows for efficient electricity control in low-voltage smart grids. Energies. Apr 2024;**17**(9):1-17. WOS: 001219947000001, e-ISSN 1996-1073, IDS Number QH3L5. DOI: 10.3390/en17092123

[9] Kundur P. Power System Stability & Control. 2nd ed. New York, United States: McGraw Hill; 2022

[10] Liserre M, Teodorescu R, Rodriguez P. Photovoltaic Inverter Structures. In the: Grid Converters for Photovoltaic and Wind Power Systems. Chichester, United Kingdom: John Wiley and Sons; 2011. pp. 26-29

[11] Pedersen A, Ahmed I, Mihet-Popa L. Chapter hierarchical control of an islanded AC micro grid using FS-MPC and an EMS. In: Mihet-Popa L, editor. Open Access Peer-Reviewed Books Smart Grids Technology and Applications. London, UK: IntechOpen; 2023. DOI: 10.5772/intechopen.110815

[12] Ciobanu I, Lazaroiu GC, Fulginei FR. Energy community part of Smart City. In: 2023 IEEE International Smart Cities Conference (ISC2). IEEE; 2023. pp. 1-7

[13] Vittal V, Ayyanar R. Grid integration and dynamic impact of wind energy. In: Power Electronics and Power Systems. New York, NY: Springer; 2013. 150p. DOI: 10.1007/978-1-4419-9323-6. Hardcover ISBN 978-1-4419-9322-9, Softcover ISBN 978-1-4899-9845-3, Published: 08 August 2014, eBook ISBN 978-1-4419-9323-6, Published: 23 July 2012, Series ISSN 2196-3185, Series E-ISSN 2196-3193, NY: State Published; 21 July 2012. 148p